THE ELEMENTS OF PROBABILITY

SIMEON M. BERMAN
New York University

THE ELEMENTS OF PROBABILITY

ADDISON-WESLEY PUBLISHING COMPANY
Reading, Massachusetts · Menlo Park, California · London · Don Mills, Ontario

The Addison-Wesley Series in
BEHAVIORAL SCIENCE: QUANTITATIVE METHODS

Frederick Mosteller, *Consulting Editor*

TO IONA

PREFACE

This is an elementary introduction to the theory and applications of probability. The only *formal* mathematical knowledge essential for an understanding of this book is that of the algebraic properties of the real number system (the "ordered field" properties) as usually presented in high school algebra. There is no attempt to teach or use calculus here. The only concept from the latter subject that is used is that of a limit of a sequence of real numbers; a self-contained exposition is given in Chapter 3.

The aim of the book is the presentation of the more *profound, interesting,* and *useful* results of classical probability in elementary mathematical forms. The pedagogic principle shaping this book is that important ideas can and should be taught at the most elementary level: that the student benefits more from a few profundities than from a lot of trivia. Cast here in primitive but meaningful forms are topics which are usually given in more advanced books: Bernoulli's Law of Large Numbers, the Glivenko-Cantelli theorem on the convergence of the empirical distribution function (Clustering Principle), the random walk, branching processes, Markov chains, and sequential testing of hypotheses. Applications and numerical illustrations complement every theoretical result of importance; for example, the coin-tossing model is applied to medical, psychological, and industrial testing and to electorate polling.

A quick examination of this book will reveal two related features:

1) It is written in the style of plane geometry: definitions and axioms are stated and their logical consequences are derived. Every result, except for the indicated few, is rigorously proved. Whenever the complete proof of a general theorem would require too much notation, we prove only a particular case, in this way simplifying but conserving the ideas of the general proof.

2) Symbolic notation is minimized: the student is forced to work in "prose."

The successive chapters strongly depend on their predecessors. The reader should check every cross-reference to definitions or propositions: the only way to understand and remember them is to see how they are used. Exercises follow each section: they consist of simple applications of stated results, questions about structures of given proofs, and simple theoretical extensions and verifications. The instructor should preserve the given proportion of theoretical to numerical material; for example, he should *not* spend most of the class time reviewing the exercises—the expository material is more important.

The first seven chapters, and the last—more than half the book—are devoted to the single model of the tossing of a coin n times; no other probability models are introduced until the eighth chapter. The fundamental notions—random variables, the Law of Large Numbers, normal approximation, events, independence—are first "grown" within the coin-tossing model; then, only after the reader is at ease in this particular case, they are extended to the more general case. While most other authors begin with set theory and the axioms of probability measures, I postpone the "calculus of events" until there is a natural demand for it—in the random-walk model. Combinatorial theory is minimized: just enough to derive the binomial distribution is given. Many other books at this level contain exercises which are applications of elementary combinatorial theory. These have been intentionally left out of this book: I have found it preferable to develop this systematic study of the foundations of probability without them.

Another novelty of the exposition is the use of capital letters in nouns representing random variables and lower-case letters in the same nouns representing values of the random variables; for example, the Number of heads in n tosses of a coin refers to a random variable. The terminology used differs from the conventional at certain points; for example, the traditional

"Bernoulli trial" is called just a "coin toss," and Cain gambles with Abel instead of Peter with Paul.

The material is annotated as follows. Each chapter, with the exception of Chapter 4, is divided into *sections*. Definitions, examples, and propositions are labeled by chapter and number; for example, Proposition 7.2 is the second in Chapter 7. The exercises are numbered only within the section; each exercise *set* is numbered. Numbered displays, of which there are not many, are labeled by chapter and number; they are usually referred to as "formulas." The tables labeled with roman numerals are located in a separate section on pages 200–205.

Here is the content of a one-semester course intended for students with some interest in axiomatic mathematics but not yet well trained in it; they may have other fields of primary specialization. Even though calculus is not a prerequisite for this book, it does contain sufficient theoretical material for those who have had calculus.

I wish to thank Professor Frederick Mosteller for helpful, constructive comments on the original manuscript; and Professor J. D. Kuelbs for additional remarks. I appreciate the permission of Holden-Day, Inc., to use Tables I and III.

September 1968   S. M. B.
Brooklyn, New York

CONTENTS

xii   CONTENTS

chapter 1

POPULATIONS, SAMPLES, AND THE BINOMIAL THEOREM

In this chapter we introduce the simplest elements of "combinatorial theory": the numbers of ordered and unordered samples from a population. These elements are used in the proof of the Binomial Theorem (Section 1.2), the derivation of the binomial distribution (Section 2.1), and the model of random sampling (Section 8.1).

1.1 ORDERED AND UNORDERED SAMPLES

Consider an urn containing a number of balls; these we call the *population*. Suppose that some of these are selected and taken out; these we call the *sample*. Samples can be distinguished one from another according to two different criteria:

1) Suppose the balls in the sample are sequentially selected from the population and that the particular *order* in which the balls are drawn is important. The first criterion considers two samples different either if they have different memberships (at least one element not in common) or if they have the same members but have been drawn in different orders. Each distinct sample is called an *ordered sample*.

2) Suppose the balls in the sample are not drawn in order; or, if they were, that the order is not important. Samples are distinguished one from another, according to the second criterion, by membership alone: two samples are distinct if they do not have exactly the same members. In this case each distinct sample is called an *unordered* sample.

Example 1.1 An urn contains five balls individually marked by the letters *A*, *B*, *C*, *D*, and *E*. A sample of three is sequentially selected and consists of the balls labeled *A*, *E*, and *D*, drawn in that order. This is a particular *ordered* sample; however, there are other ordered samples which are indistinguishable from it according to the second criterion. The ordered sample of balls labeled *A*, *D*, and *E* (in that order) has the same members; therefore, it is the same *unordered* sample as the first one. There are several other ordered samples indistinguishable by the second criterion from these two: *D, A, E*; *D, E, A*; *E, D, A*; and *E, A, D*.   ◁

In the above example we saw that three balls can be arranged in six different orders. This can be reaffirmed by means of a simple principle. We can think of the arrangements of the three balls as the seating orders of three people in three chairs. The first chair can be taken by any one of three persons; the second can be taken by either of the remaining two; and the third taken by the single remaining one. Therefore, the three can be seated in $3 \cdot 2 \cdot 1 = 6$ possible ways. We shall use this principle to derive the formula for the number of distinct ordered samples that can be drawn from a population of a given size. The letters N and s are used to represent the numbers of balls in the population and in the sample, respectively.

Proposition 1.1 *The number of ordered samples of size s that can be selected from a population of size N is*

$$N(N - 1) \cdots (N - s + 1), \qquad (1.1)$$

where the product contains s factors.

PROOF. Let us first present the proof for the particular case $s = 2$, an ordered sample of size 2.

In the formation of an ordered sample there are N balls eligible to be drawn first; hence, there are N ways of filling the first "slot" in the ordered sample. After the first ball has been selected, there are $N - 1$ remaining balls eligible to be drawn for the second slot; hence, the latter can be filled in $N - 1$ ways. Let the letters $A_1, \ldots, A_N$ be the labels on the balls. Each ordered sample of two balls is put into correspondence with exactly one of the following pairs of letters:

$$A_1A_2, \; A_1A_3, \; \ldots, \; A_1A_N$$
$$A_2A_1, \; A_2A_3, \; \ldots, \; A_2A_N$$
$$\vdots \qquad \vdots$$
$$A_NA_1, \; A_NA_2, \; \ldots, \; A_NA_{N-1}$$

For example, the ordered sample consisting of A_1 and A_3 on the first and second draws, respectively, corresponds to the pair A_1A_3. The above display has N rows and $N - 1$ columns; thus, there are $N(N - 1)$ displayed pairs of letters; therefore, the number of ordered samples of size $s = 2$ is $N(N - 1)$.　　　　　◀

The argument above can be extended to a general number s of balls in the sample. The first two slots can be filled in $N(N - 1)$ ways. The third ball is chosen from the remaining $N - 2$; hence, there are $N - 2$ ways of filling the third slot; therefore, by extending the display of *pairs* of letters above to one of triples, we find that there are $N(N - 1)(N - 2)$ ways of filling the first three slots. Formula (1.1) is obtained by successive application of this principle to the remaining slots.

Example 1.2 Let us illustrate the proof of Proposition 1.1 in the particular case of a population of four balls A, B, C, and D and ordered samples of three. The first slot can be filled in four ways, the second in three, and the third in two. The samples are

$$
\begin{array}{cccc}
ABC & BAC & CAB & DAB \\
ACB & BCA & CBA & DBA \\
ADC & BDC & CAD & DCB \\
ACD & BCD & CDA & DBC \\
ABD & BDA & CBD & DCA \\
ADB & BAD & CDB & DAC
\end{array}
$$

The number of ordered samples is 24, which, in accordance with formula (1.1) for $N = 4$ and $s = 3$, is $4 \cdot 3 \cdot 2 = 24$.　　　　◁

There is a relation between the number of ordered and unordered samples of a given size.

Example 1.3 Let us find all ordered samples in the previous example consisting of the balls A, B, and C; they are

$$ABC, \ ACB, \ BAC, \ BCA, \ CAB, \ CBA.$$

There are six of these, as in Example 1.1; thus, the six ordered samples are the same unordered sample; and, furthermore, every *unordered* sample of three can be arranged as six distinct ordered samples. It follows that there are six times as many ordered as unordered samples. The four unordered samples are ABC, ABD, ACD, and CBD. $\triangleleft$

This example illustrates the following two propositions.

Proposition 1.2 *Each unordered sample of s balls can be arranged in* $1 \cdot 2 \cdots s$ *different ordered samples.*

PROOF. We employ the same principle as that used for Proposition 1.1. (The present proposition is really a consequence of the latter, but the proof is repeated to assist the reader.) The first slot for the ordered sample of s balls can be filled from among the s given balls; the second slot in $s - 1$ ways from among the remaining balls; the third slot in $s - 2$ ways from among the remaining ones, etc.; hence, the ordered sample can be formed in $s(s - 1) \cdots 1$ ways. This product of s numbers is equal to that stated in the proposition.   ◀

The product $1 \cdot 2 \cdots s$ of the first s integers is universally denoted as $s!$ (pronounced "s factorial").

Proposition 1.3 *The number of unordered samples of size s from a population of size N is*

$$N(N - 1) \cdots (N - s + 1)/s!. \ = \ \binom{N}{S} \qquad (1.2)$$

PROOF. Note that the numerator in formula (1.2) is the number of *ordered* samples of s from a population of N (Proposition 1.1) and that the denominator is the number of ordered samples of s derived from a particular unordered sample (Proposition 1.2). The proposition follows from these two remarks and from the fact:

[number of ordered samples] = [number of ordered
samples formed from a single unordered sample]
$\times$ [number of unordered samples].   ◀

1.1 EXERCISES

1. An urn contains five balls marked A, B, C, D, and E. Enumerate the ordered samples of size 2. How many ordered samples can be derived from each unordered one?

2. An urn contains six balls marked A, B, C, D, E, and F. Enumerate the unordered samples of size three. How many ordered samples can be derived from each unordered one, and how many ordered samples are there?

3. An urn contains 100 marked balls. How many ordered and unordered samples of size 4 are there?

4. Find the numbers of ordered and unordered samples of s from the populations of size N in each of these cases:

a) $N = 6$, $s = 2, 4, 5$. b) $N = 8$, $s = 4, 7$.
c) $N = 10$, $s = 4, 6$.

1.2 THE BINOMIAL THEOREM

We shall use Proposition 1.3 to give a proof of the Binomial Theorem of algebra. This is the rule that permits the expression of any power of a binomial as a sum of products of powers of the components. In raising the binomial $A + B$ to successive integral powers, we find

$$(A + B)^1 = A + B,$$
$$(A + B)^2 = A^2 + 2AB + B^2,$$
$$(A + B)^3 = A^3 + 3A^2B + 3AB^2 + B^3,$$
$$(A + B)^4 = A^4 + 4A^3B + 6A^2B^2 + 4AB^3 + B^4.$$

In each of these examples, the terms of the expansion of $(A + B)^N$ are obtained in accordance with the following rules:

1. The first term is A^N, $N = 1, 2, 3, 4$.

2. In each successive term, the exponent of A decreases by 1 and the exponent of B increases by 1.

3. The numerical coefficient of the term containing B^s in the expansion of $(A + B)^N$ is given by formula (1.2) (with the apparent exception of the first and last terms); for example, the coefficient of AB^3 is $4 \cdot 3 \cdot 2/3 \cdot 2 \cdot 1 = 4$.

Let us justify these rules in the particular case $N = 3$. The product $(A + B)(A + B)$ has four single terms $AA + AB + BA + BB$; and $(A + B)^3 = (A + B)(A + B)(A + B)$ has eight terms

$$AAA + AAB + ABA + BAA + BBA + BAB + ABB + BBB.$$

There is one term containing three A-factors, and one with three B-factors; therefore, A^3 and B^3 appear with coefficient 1. The number of terms with exactly one B-factor (and two A-factors) is the number of different ways that the letter B can be placed into one of three slots, namely, three. The number of terms with exactly two B-factors (and one A-factor) is the number of different ways that two letters B can be arranged in three slots. This is analogous to the number of ways that an unordered sample of two balls can be selected from a population of three: each of the three slots represents a ball in the population, and each slot filled with a letter B represents a ball selected for the sample; therefore, the number of terms with two B-factors is given by formula (1.2) with $s = 2$ and $N = 3$: $3 \cdot 2/2 \cdot 1 = 3$.

These considerations will be generalized. Formula (1.2) is conventionally denoted by the symbol

$$\binom{N}{s} = N(N - 1) \cdots (N - s + 1)/s! \tag{1.3}$$

for integers s assuming values from 1 to $N - 1$. For convenience, we also define the symbol for $s = 0$ and $s = N$:

$$\binom{N}{0} = \binom{N}{N} = 1. \tag{1.4}$$

The numbers given in formulas (1.3) and (1.4) are known as the "binomial coefficients" because of their positions in the

Binomial Formula. $(A + B)^N$ *is equal to*

$$\binom{N}{0} A^N + \binom{N}{1} A^{N-1} B + \cdots + \binom{N}{s} A^{N-s} B^s + \cdots + \binom{N}{N} B^N. \tag{1.5}$$

PROOF. This proof (as well as most of the subsequent ones) is broken up into a series of numbered statements, each followed by a supporting reason.

1. $(A + B)^N$ is equal to the sum of all products of N factors formed from the two letters A and B. *Distinct* products are formed by factors in *different* orders.

Reason. This follows from the very definition of multiplication of binomials; it is illustrated above in the case $N = 3$.

2. The number of products (in Statement 1) containing exactly s factors B and $N - s$ factors A is equal to the number of unordered samples of s balls that can be drawn from a population of N.

Reason. Each of the N slots in a product is like a ball in a population, and each of the slots filled with a factor B is like a ball selected for the population; hence, the number of ways of forming such a product containing exactly s factors B is the number of ways of forming an unordered sample of size s.

3. The assertion of the proposition follows from Statements 1 and 2.

Reason. In writing the sum of all products in Statement 1, one may write the sum of all terms containing a *common number* of A-factors as a single term multiplied by the *number* of such terms; hence, the sum of all such products can be represented in the form (1.5). ◀

Example 1.4 The binomial coefficients for $N = 5$ are:

$$\binom{5}{0} = \binom{5}{5} = 1; \quad \binom{5}{1} = 5; \quad \binom{5}{2} = \frac{5 \cdot 4}{2 \cdot 1} = 10;$$

$$\binom{5}{3} = \frac{5 \cdot 4 \cdot 3}{3 \cdot 2 \cdot 1} = 10; \quad \binom{5}{4} = \frac{5 \cdot 4 \cdot 3 \cdot 2}{4 \cdot 3 \cdot 2 \cdot 1} = 5;$$

therefore, the binomial expansion is

$$(A + B)^5 = A^5 + 5A^4B + 10A^3B^2 + 10A^2B^3 + 5AB^4 + B^5. \quad ◁$$

Note the symmetry of the coefficients in the latter binomial expansion as well as in the previous ones: the coefficient of $A^{N-s}B^s$ is equal to the coefficient of A^sB^{N-s}. This is due to the symmetry of the positions of A and B in the expression $(A + B)^N$: the latter is equal to $(B + A)^N$; hence, the coefficient of a particular power of A in the expression for $(A + B)^N$ is identical with the coefficient of the same power of B in the (equivalent) expression for $(B + A)^N$:

$$\binom{N}{s} = \binom{N}{N - s}. \tag{1.6}$$

This equation has the following interpretation in terms of unordered samples: each unordered sample of s balls *taken from* an urn containing N balls is associated with exactly one unordered sample of $N - s$ balls formed from those *left in* the urn; hence, for each one of the former there is exactly one of the latter. Equation (1.6) states that there are as many unordered samples of s balls as there are of $N - s$ balls.

The binomial coefficients have other interesting properties. We remark that the factorial expression $s!$ satisfies the equation

$$s! = s \cdot (s - 1)!; \tag{1.7}$$

for example, 7! is equal to $7 \cdot 6!$. The reason for the general equation (1.7) is that the right-hand side is equal to the product of the first $s - 1$ integers multiplied by s, which is identical with the left-hand side. Now we prove the validity of the following formula:

$$\binom{N-1}{s-1} = \frac{s}{N}\binom{N}{s}. \tag{1.8}$$

The right-hand side of Eq. (1.8) is, by formula (1.3), equal to

$$\frac{s}{N} \cdot \frac{N(N-1)\cdots(N-s+1)}{s!}.$$

Apply Eq. (1.7) to the denominator, and then cancel the factors s and N from the numerator and denominator; the fraction remaining is

$$\frac{(N-1)\cdots(N-s+1)}{(s-1)!} = \frac{(N-1)\cdots(N-1-(s-1)+1)}{(s-1)!}$$

$$= \binom{N-1}{s-1}.$$

The last member is on the left-hand side of Eq. (1.8).

Example 1.5 Let us show that $\binom{5}{3}$ is equal to $\frac{4}{6} \cdot \binom{6}{4}$. The latter is

$$\frac{4}{6} \cdot \frac{6 \cdot 5 \cdot 4 \cdot 3}{4 \cdot 3 \cdot 2 \cdot 1} = \frac{5 \cdot 4 \cdot 3}{3 \cdot 2 \cdot 1} = \binom{5}{3}. \qquad \triangleleft$$

1.2 EXERCISES

1. Give a detailed proof of the Binomial Formula in the particular cases $N = 4, 5$. Write down all the products of factors A and B, and show that their sum is that given by the formula.

2. Find all the binomial coefficients in the cases $N = 7, 8$.

3. Find the coefficient of $A^{10}B^4$ in the binomial expansion of $(A + B)^{14}$; the coefficient of A^6B^5 in the expansion of $(A + B)^{11}$; and the coefficient of A^8B^3 in the same expansion.

4. Show that $\binom{7}{4}$ is equal to $\frac{5}{8} \cdot \binom{8}{5}$; and $\binom{10}{3}$ is equal to $\frac{4}{11} \cdot \binom{11}{4}$.

5. Prove the validity of the equation

$$\binom{N+1}{s} = \binom{N}{s-1} + \binom{N}{s}$$

by substitution in formula (1.2) and application of Eq. (1.7).

6. Extend Eq. (1.8) to

$$\binom{N-2}{s-2} = \frac{(s-1)s}{(N-1)N}\binom{N}{s}$$

by multiplying each side of Eq. (1.8) by $(s-1)/(N-1)$, and then applying Eq. (1.8) itself with $N-2$, $N-1$, $s-2$, and $s-1$ in place of $N-1$, N, $s-1$, and s, respectively.

7. Show that, for a fixed s, $N^s/s!$ is a good approximation to $\binom{N}{s}$ if N is very large; examine the ratio of the two quantities. Estimate the latter ratio in the particular case $N = 100$, $s = 3$.

chapter 2

THE COIN-TOSSING GAME

A simple model illustrating the fundamental ideas of probability theory is the coin-tossing game. In this chapter we first informally describe the mathematics of a game of one, two, and three tosses; and then list the formal definitions and axioms defining a game of n tosses, where n is an arbitrary positive integer. It is shown that the binomial distribution gives the probabilities of various numbers of heads. In the second section the theory of the coin game is applied to problems arising in other fields.

2.1 THEORY

A player tosses a coin and observes the result, which is either "head" or "tail." We shall use the letters H and T to represent these outcomes. It is impossible before the toss to foretell with certainty whether H or T will turn up: each is a *possible* outcome for the toss. The *actual* outcome that turns up as a result of the toss is a *variable*. It assumes exactly one of the possible outcomes H and T; we shall call it the Outcome (with a capital "O") of the toss. With the outcome H we associate a positive real number called the *probability* of H; and with the outcome T a positive real number called the *probability* of T. These probabilities are mathematical quantities assumed as part of the coin-tossing model; they are taken to have the property that their sum is 1. The probabilities of H and T may be interpreted as measures of "likelihood" of the occurrence of H and T, respectively; for example, if the probabilities of H and T are $\frac{3}{4}$ and $\frac{1}{4}$, respectively, a rational gambler would surely bet that H, not T, will turn up on a toss of the coin. The coin is said to be "balanced" or "fair" if these probabilities are equal (and necessarily equal to $\frac{1}{2}$).

If the coin is tossed twice, the possible outcomes are represented as

$$[HH], \quad [HT], \quad [TH], \quad [TT], \tag{2.1}$$

where [HH] means that H occurs on both tosses, [HT] means that H occurs on the first toss and T on the second toss, etc. The Outcome of the *pair* of tosses is a variable which assumes the form of any of the four pairs in (2.1). Suppose that the Outcome of the first toss does not affect that of the second, nor is affected by it; in this case, when the coin is also balanced, each of the four outcomes in (2.1) is assigned probability $\frac{1}{4}$.

The *number* of H's observed as a result of the two tosses is a variable which assumes exactly one of the values 0, 1, and 2: the outcome [HH] yields two H's, the outcomes [HT] and [TH] yield one H each, and the outcome [TT] none. This variable quantity will be called the Number of H's (with a capital "N") in two tosses. We define the probability that the Number of H's is 0 as the probability of the corresponding outcome [TT], which is $\frac{1}{4}$. The probability that the Number of H's is equal to 1 is defined as the sum of the probabilities assigned to the corresponding outcomes [HT] and [TH], namely, $\frac{1}{4} + \frac{1}{4} = \frac{1}{2}$. Finally, the probability that the Number of H's is equal to 2 is the probability of the outcome [HH], which is $\frac{1}{4}$. The system of probabilities $\frac{1}{4}, \frac{1}{2}, \frac{1}{4}$ for the Number of H's is called the "probability distribution" of the Number of H's.

Suppose the coin is not necessarily balanced, that is, that the probabilities of H and T are not necessarily equal to $\frac{1}{2}$; for example, suppose that the coin is shaven down on the head side so that it is "more likely" to fall with that side up. We denote by a number p, where p is between 0 and 1, the measure of how "likely" H is to appear on a toss: the closer p is to 1, the more likely it is to appear. If the coin is shaven on the head side, then p exceeds $\frac{1}{2}$; if shaven on the tail side, it is less than $\frac{1}{2}$. We put $q = 1 - p$; then q is a measure of the likelihood of T. We call p and q the probabilities that H and T appear, respectively. In the case of the balanced coin p and q are both equal to $\frac{1}{2}$. The numbers p and q are not observable quantities but are mathematical creatures in a system of axioms; however, it will be shown that p and q can be estimated by observing the outcomes of many tosses of the coin (Law of Large Numbers, Chapter 3). All the definitions and propositions below are valid for any choice of probabilities p and q.

If the coin is tossed twice, the possible outcomes are again those in the display (2.1); however, the probabilities are not necessarily equal (to $\frac{1}{4}$). We assume as before that the Outcomes of the two tosses have no mutual effect; then we assign the probabilities in the following way:

$$
\begin{array}{ccccc}
\text{outcome:} & [HH] & [HT] & [TH] & [TT] \\
\text{probability:} & p^2 & pq & qp & q^2
\end{array}
\qquad (2.2)
$$

For example, the Outcome is equal to [TH] with probability qp. These probabilities are positive because p and q are both positive. The sum of the four probabilities is equal to 1 because $p + q$ is 1:

$$ p^2 + pq + qp + q^2 = (p + q)^2 = 1^2 = 1. $$

The Number of H's in the two tosses may be 0, 1, and 2; however, the probabilities are no longer necessarily equal to $\frac{1}{4}$, $\frac{1}{2}$, and $\frac{1}{4}$, respectively. The probability that the Number of H's is equal to 0 is defined as q^2, the probability that the Outcome is [TT]. The probability that the Number of H's is 1 is defined as the sum of the probabilities that the Outcome is [HT] and [TH], namely $2pq$. The probability that the Number of H's is 2 is defined as the probability that the Outcome is [HH], that is, p^2. The system of probabilities q^2, $2pq$, and p^2 is called the "probability distribution" of the Number of H's in two tosses of the coin.

Suppose the coin is tossed three times, and that there are no mutual effects among the tosses. The set of possible outcomes and their corresponding probabilities is:

$$
\begin{array}{llll}
\text{outcome} & \text{probability} & \text{outcome} & \text{probability} \\
[HHH] & p^3 & [TTT] & q^3 \\
[HHT] & p^2q & [TTH] & q^2p \\
[HTH] & p^2q & [THT] & q^2p \\
[THH] & p^2q & [HTT] & q^2p
\end{array}
\qquad (2.3)
$$

Each outcome is evidently assigned a probability which is a product of the factors p or q or both: there is a factor p for each H appearing in the outcome and a factor q for each T. The probabilities are all positive because p and q are. The sum of the probabilities is 1:

$$ p^3 + 3p^2q + 3pq^2 + q^3 = (p + q)^3 = 1^3 = 1. $$

The probability that the Number of H's in three tosses is 0 is defined as the probability that the Outcome is [TTT]: this is q^3. The probability that the Number of H's is 1 is defined as the sum of the probabilities that the Outcome is [TTH], [THT], and [HTT], respectively: it is $3q^2p$. In the same way we define the probabilities that the Number of H's is 2 and 3 as $3qp^2$ and p^3, respectively. It follows that the probability distribution of the Number of H's is:

$$
\begin{array}{lccccc}
\text{Number of H's:} & 0 & 1 & 2 & 3 & \\
\text{probability:} & q^3 & 3q^2p & 3qp^2 & p^3 & (2.4)
\end{array}
$$

The above description of the coin-tossing game is now generalized to the case of an *arbitrary* number of tosses. We shall use the letter n for the number of tosses. We start with three formal definitions.

Definition 2.1 A coin is tossed n times. The system of outcomes of the tosses is the sytem of all multiplets of n letters that can be formed from the two letters H and T. The letter H (or T) is put in the first, second, . . . slot in the multiplet when H (or T) turns up on first, second, . . . toss, respectively.

Example 2.1 A coin is tossed 5 times; here $n = 5$. The multiplet [THHTH] is the outcome: H appears on the second, third, and fifth tosses and T appears on the first and fourth tosses. ◁

Definition 2.2 The probability that the Outcome is equal to a particular multiplet is a product of factors p or q or both. There are n factors in the product: the numbers of factors p is the number of H's in the multiplet, and the number of factors q is the number of T's.

Example 2.2 In five tosses, the probability that the Outcome is [THHTH] is p^3q^2. ◁

Definition 2.3 The probability that the Number of H's in n tosses is a specified number k is defined as the sum of the probabilities of all outcomes whose corresponding multiplets contain exactly k letters H.

Example 2.3 The probability that the Number of H's in five tosses is 1 is the sum of the probabilities of the outcomes [HTTTT], [THTTT], [TTHTT], [TTTHT], and [TTTTH]. By Definition 2.3, each of these has probability pq^4; hence, the sum of the probabilities is $5pq^4$. ◁

Proposition 2.1 *In the system of multiplets described in Definition 2.1, there are exactly* $\binom{n}{k}$ *consisting of k letters H and n − k letters T.*

PROOF. The proof is almost the same as that of the Binomial Formula (Section 1.2) but we present it here for completeness.

1. The number of multiplets containing exactly k letters H and $n - k$ letters T is equal to the number of unordered samples of k balls from a population of n.

Reason. The n slots in the multiplet are like n balls in an urn; and the slots containing H's are like the balls selected for the sample. The number of ways of filling exactly k slots with H's is the number of ways of forming an unordered sample of k balls.

2. The number of unordered samples of k balls from a population of n is $\binom{n}{k}$.

Reason. Proposition 1.3, and formula (1.3): substitute n for N and k for s in the expression $N(N - 1) \cdots (N - s + 1)/s!$.   ◄

Example 2.4 Let us find the number of multiplets of six letters formed from H and T containing exactly four letters H. According to Proposition 2.1 (with $n = 6$, $k = 4$) the number is

$$\binom{6}{4} = \frac{6 \cdot 5 \cdot 4 \cdot 3}{4 \cdot 3 \cdot 2 \cdot 1} = 15.$$

This can also be verified by direct enumeration; the 15 multiplets are:

[HHHHTT]	[HHHTTH]	[HTHTHH]
[HHHTHT]	[HHTHTH]	[THHTHH]
[HHTHHT]	[HTHHTH]	[HTTHHH]
[HTHHHT]	[THHHTH]	[THTHHH]
[THHHHT]	[HHTTHH]	[TTHHHH]

◁

Example 2.5 The number of multiplets of 10 letters formed from H and T which contain exactly three letters H is (with $n = 10$, $k = 3$)

$$\binom{10}{3} = \frac{10 \cdot 9 \cdot 8}{3 \cdot 2 \cdot 1} = 120.$$

◁

Proposition 2.1 is used in finding the probability distribution of the Number of H's in several tosses of the coin. First we present two examples and then state and prove the general result.

Example 2.6 Find the probability that the Number of H's in six tosses is 4. In Example 2.4 it was shown that there are 15 multiplets containing exactly four letters H. By Definition 2.2 each of these has probability p^4q^2. Finally, by Definition 2.3, the probability that the Number of H's is 4 is the sum of the 15 (equal) probabilities p^4q^2; therefore, it is $15p^4q^2$.   ◁

Example 2.7 Find the probability that the Number of H's in 10 tosses is 3. In Example 2.5 it was shown that there are 120 multiplets containing exactly three letters H. By Definition 2.2 each of these has probability p^3q^7. Finally, by Definition 2.3, the probability that the Number of H's is 3 is the sum of the 120 (equal) probabilities p^3q^7; therefore, it is $120p^3q^7$.   ◁

Here is the general principle:

Proposition 2.2 *The probability of any particular outcome of n tosses of a coin containing exactly k letters H (and n − k letters T) is p^kq^{n-k}.*

PROOF. Definitions 2.1 and 2.2.   ◀

Proposition 2.3 *The probability that the Number of H's in n tosses of the coin is equal to k is given by the formula*

$$\binom{n}{k} p^k q^{n-k}, \qquad k = 0, 1, \ldots, n. \tag{2.5}$$

PROOF

1. Each outcome with k letters H has probability p^kq^{n-k}.

Reason. Proposition 2.2.

2. There are $\binom{n}{k}$ outcomes with k letters H.

Reason. Proposition 2.1.

The assertion of the proposition follows from these two statements and Definition 2.3.   ◀

The system (2.5) of $n + 1$ probabilities is called the *binomial distribution*. It is tabulated for selected values of n and p (Table I). Graphs of various binomial distributions with $n = 5$ are given in Fig. 2.1. The black triangles touch the horizontal axes at the points np. The content of Proposition 2.3 is that this is the probability distribution of the Number of H's appearing in n tosses of a coin. These probabilities are positive because p and q are.

Probability
of *k* H's

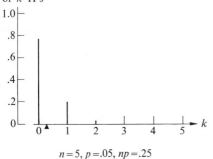

$n = 5, p = .05, np = .25$

(a)

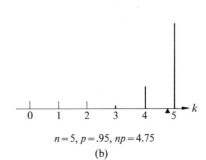

$n = 5, p = .95, np = 4.75$

(b)

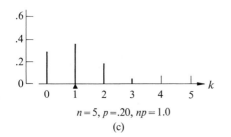

$n = 5, p = .20, np = 1.0$

(c)

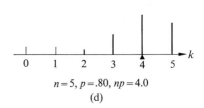

$n = 5, p = .80, np = 4.0$

(d)

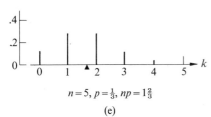

$n = 5, p = \frac{1}{3}, np = 1\frac{2}{3}$

(e)

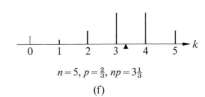

$n = 5, p = \frac{2}{3}, np = 3\frac{1}{3}$

(f)

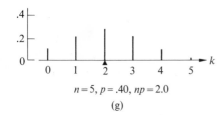

$n = 5, p = .40, np = 2.0$

(g)

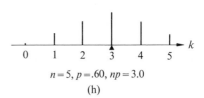

$n = 5, p = .60, np = 3.0$

(h)

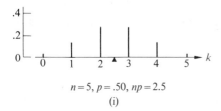

$n = 5, p = .50, np = 2.5$

(i)

Fig. 2.1. Binomial distributions for $n = 5$, displaying the change in form as p varies. (Adapted from Frederick Mosteller, Robert E. K. Rourke, and George B. Thomas, Jr., *Probability with Statistical Applications*, Reading, Mass: Addison-Wesley, 1961.)

The sum of the probabilities is 1 by virtue of the Binomial Formula:

$$\binom{n}{0} q^n + \binom{n}{1} q^{n-1} p + \cdots + \binom{n}{n} p^n = (q + p)^n = 1^n = 1. \quad (2.6)$$

The sum in the first member of this equation is the sum of the terms of the binomial distribution (2.5).

Example 2.8 A coin with $p = \frac{1}{2}$ is tossed $n = 5$ times. Find the probability that the Number of H's is equal to 2. Put $p = q = \frac{1}{2}$, $n = 5$, and $k = 2$ in expression (2.5):

$$\binom{5}{2} \left(\frac{1}{2}\right)^2 \left(\frac{1}{2}\right)^3 = \frac{5 \cdot 4}{1 \cdot 2} \left(\frac{1}{2}\right)^5 = \frac{10}{32} = \frac{5}{16}. \qquad \triangleleft$$

Example 2.9 A coin with probability $p = \frac{1}{3}$ is tossed $n = 4$ times. Find the probability that the Number of H's is 3. Put $p = \frac{1}{3}$, $q = \frac{2}{3}$, $n = 4$, and $k = 3$ in expression (2.5):

$$\binom{4}{3} \left(\frac{1}{3}\right)^3 \left(\frac{2}{3}\right)^1 = \frac{4 \cdot 3 \cdot 2}{3 \cdot 2 \cdot 1} \frac{2}{3 \cdot 3 \cdot 3 \cdot 3} = \frac{8}{81}. \qquad \triangleleft$$

Formula (2.5) furnishes the probability that the Number of H's is equal to a *specific* integer. Now we define the probability that the Number of H's is equal to *some integer in a specific set of integers.*

Definition 2.4 The probability that the Number of H's is *one of a specific set of integers* is the sum of the probabilities in formula (2.5) whose indices k belong to that specific set.

Example 2.10 A coin is tossed four times. By the "probability that the Number of H's is *less than* 2" we mean the probability that the Number of H's is one of the set of integers 0 and 1; by Definition 2.4 this probability is the sum of the probabilities of the binomial distribution ($n = 4$) whose terms have $k = 0$ and $k = 1$: $q^4 + 4q^3 p$. The "probability that the Number of H's is *less than or equal to* 2" is taken to be the probability that the Number of H's is one of the numbers 0, 1, and 2; by Definition 2.4 this probability is the sum of the probabilities of the binomial distribution ($n = 4$) whose terms have $k = 0$, $k = 1$, and $k = 2$:

$$q^4 + 4pq^3 + 6p^2 q^2.$$

The "probability that the Number of H's is *at least* 2" is defined as the probability that the Number of H's is one of the integers 2, 3, and 4; by Definition 2.4 this probability is the sum of the probabilities of the binomial distribution ($n = 4$) whose terms have $k = 2$, $k = 3$, and $k = 4$:

$$6p^2q^2 + 4p^3q + p^4.$$

The "probability that the Number of H's is *greater than* 2" is taken as the probability that the Number of H's is one of the numbers 3 and 4; therefore, the probability is $4p^3q + p^4$. The "probability that the Number of H's is *between* 1 *and* 3 *inclusive*" is meant to be the probability that the Number of H's is one of the numbers 1, 2, and 3; it is

$$4pq^3 + 6p^2q^2 + 4p^3q. \qquad \triangleleft$$

For purposes of calculation it is often convenient to use the fact that the sum of the probabilities in the binomial distribution is 1 (Eq. 2.6).

Example 2.11 A coin with $p = .8$ is tossed 10 times. Find the probability that the Number of H's is at least equal to 2. By Definition 2.4 the probability is given by the sum of the terms of the binomial distribution whose indices run from $k = 2$ to $k = 10$:

$$\binom{10}{2}(.8)^2(.2)^8 + \cdots + \binom{10}{10}(.8)^{10}(.2)^0.$$

By Eq. (2.6) the latter sum is equal to 1 minus the sum of the terms of indices $k = 0$ and $k = 1$:

$$1 - (.2)^{10} - 10(.8)(.2)^9.$$

The latter is much easier to compute than the former sum. $\qquad \triangleleft$

We conclude this section with a remark about an assumption we made about the successive tosses of the coin before the formal definitions were given: it was assumed that the Outcomes of different tosses have no "mutual effects." This is an assumption which is not of a mathematical nature but rather of an empirical nature; for this reason it was not included in the formal mathematical definitions. The purpose of the *theory* of the coin-tossing game is to provide us with information about real coin-tossing games, or natural processes like them. It has been shown in practice that the above theory furnishes correct and useful information about *real* coin-tossing games satisfying the empirical hypothesis of no "mutual effects" among different tosses.

2.1 EXERCISES

1. Enumerate the system of outcomes for $n = 4$ tosses, and the corresponding probabilities in terms of p and q. Derive the probability distribution of the Number of H's in this case.

2. Repeat the previous exercise for $n = 5$.

3. What is the relation between the multiplets representing the outcomes of four tosses, and the individual terms in the sum obtained by expanding $(A + B)^4$? Generalize this to n tosses.

4. Find the probability of the outcome [HHHTTT] for six tosses. How many outcomes have the same number (three) of H's?

5. Find the *total* number of outcomes for six tosses; eight tosses; 10 tosses.

6. Find the probability that the Number of H's in six tosses is at least four when $p = .4$.

7. A balanced coin is tossed eight times. What is the probability that the Number of H's is four (exactly half the number of tosses)? What is the probability that the Number of H's in 12 tosses is six? Criticize the statement: "If a balanced coin is tossed several times, the number of heads and tails are about equal."

8. Find the probability that the Number of H's is at most eight when a balanced coin is tossed 10 times.

9. Find the probability that the Number of H's in eight tosses is between two and five, inclusive, when $p = \frac{1}{5}$.

10. Enumerate the terms of the binomial distribution (2.5) for $n = 7$; $n = 9$.

11. Show that the kth term in (2.5) is unchanged when p and k are interchanged with q and $n - k$, respectively.

2.2 APPLICATIONS

In this section we shall show that certain processes arising in natural and behavioral science can be considered as coin-tossing games and that the foregoing theory can be used to analyze them.

A. The Sex of a Newborn Child

Whether a child is male or female is determined at conception by a chromosome of the father: if an X-chromosome is passed to the child, it is female; and if a Y-chromosome is passed, it is male. Since equal numbers of these two kinds are available, and since there is apparently no significant preference in nature for one kind of chromosome or the other, we suppose that the selection of X or Y is like the tossing of a balanced coin: X and Y correspond to the outcomes H and T, respectively. Now suppose that there are several children in a family: successive conceptions are supposed to be independent of each other so that the Outcomes (male or female) have no mutual effects. We assume also that the number of children in a family is uninfluenced by the composition of the family; for example, parents do not necessarily cease to have children after a fourth female is born. The previous model for several tosses of a coin is applicable: the number of females is like the Number of H's in several tosses.

Suppose that there are five children in a family: what is the probability that at least two of them are males? This is the probability that the Number of T's in five tosses of a balanced coin is at least 2. According to Definition 2.4 the probability is

$$\binom{5}{2}\left(\frac{1}{2}\right)^2\left(\frac{1}{2}\right)^3 + \binom{5}{3}\left(\frac{1}{2}\right)^3\left(\frac{1}{2}\right)^2 + \binom{5}{4}\left(\frac{1}{2}\right)^4\left(\frac{1}{2}\right) + \binom{5}{5}\left(\frac{1}{2}\right)^5,$$

which is equal to $\frac{13}{16}$.

B. Drug Testing

Suppose that a certain proportion p_0 of all patients with a disease naturally recover; however, it is impossible to predict *which* patients will recover. With each patient we associate two outcomes: "recovery" or "no recovery." The fate of a patient can be imagined to be like the Outcome of the toss of a coin: H corresponds to "recovery" and T to "no recovery." We assign H probability p_0 and T probability $q_0 = 1 - p_0$.

Suppose that a group of n patients is observed. We assume that there is no interaction among these: the recovery of one is without influence on the health of the others. The number of patients that actually recover is like the Number of H's in n tosses of the coin with probability p_0 of H.

A medical research worker wants to test the effectiveness of a new drug in treating the patients. He sets a tentative hypothesis that a larger proportion p_1 would recover if all patients were so treated. Suppose that he gives the drug to an experimental group of n patients, and, furthermore, that the recovery of a patient in the experimental group is also without influence on others. Under the hypothesis that the proportion of recovery is p_1, the number of patients that actually recover is like the Number of H's in n tosses of a coin with probability p_1 of H.

Here is an example. Suppose that the natural recovery proportion for a disease is $p_0 = .4$. A proposed drug is supposed to raise the recovery proportion to $p_1 = .8$. An experimental group of four patients is given the drug. Here are the probabilities for the number of cured patients in the two cases $p_0 = .4$ and $p_1 = .8$.:

number of recoveries	probabilities	
	$p_0 = .4$	$p_1 = .8$
0	.1296	.0016
1	.3456	.0256
2	.3456	.1536
3	.1536	.4096
4	.0256	.4096

These probabilities can be used to test the hypothesis that the drug really raises the recovery rate to .8 against the alternative hypothesis that it is really worthless and the recovery rate in the experimental group is still .4. The probability of at least three recoveries is the probability of at least three H's in four tosses: by Definition 2.4 it is $.1536 + .0256 = .1792$ when $p = .4$, and $.4096 + .4096 = .8192$ when $p = .8$; thus, it is about "four times more likely" to happen when p_1 is the recovery rate than when p_0 is. The probability of at most two recoveries is, by analogous reasoning, equal to $.1296 + .3456 + .3456 = .8208$ when $p = .4$, and $.0016 + .0256 + .1536 = .1808$ when $p = .8$; therefore, it is about "four times more likely" to happen when p_0 is the recovery rate than when p_1 is. We feel that if at least three recoveries are observed, then $p_1 = .8$ is more likely than $p_0 = .4$ to be the real recovery rate in the group; on the other hand, if there are at most two recoveries, then $p_0 = .4$ is more likely to be the real recovery rate. In the former case, we conclude that the drug is effective, and in the latter case that the drug is not effective.

C. Psychological Testing for Extrasensory Perception (ESP)

Psychologists have used the following experiment to get evidence for the existence of ESP. An experimenter and a subject play a "card-calling" game with a deck of 25 cards consisting of five kinds of cards, five of a kind. The experimenter first shuffles the cards, then draws them one at a time. The subject does not see the cards. Before each card is drawn he states a guess as to what *kind* of card will be drawn. The experimenter marks each guess as correct or incorrect, and, after the experiment, counts the number of correct guesses among the 25. Under the hypothesis that a subject does not possess powers of ESP, his guesses are made "at random": he has "one chance in five" of being correct on each, and the correctness of the successive guesses are without mutual influence. It follows that the guesses are like the Outcomes of 25 tosses of a coin with $p = \frac{1}{5}$: H and T correspond to correct and incorrect guesses, respectively. Under the alternative hypothesis that the subject has powers of ESP, he has probability greater than $\frac{1}{5}$ of correctly guessing individual cards; however, he is not expected to correctly guess *every* card. The following criterion is used to prove the existence of ESP in a subject. If the probability, under the assumption of "random guessing" ($p = \frac{1}{5}$, no ESP), of getting more than a certain number of correct guesses is extremely small, and if a subject *does* correctly guess more than that number, then we should reject the assumption $p = \frac{1}{5}$ in favor of the alternative that p is really greater than $\frac{1}{5}$.

Here is an example. The probability of more than 10 correct guesses out of 25 is, according to Definition 2.4, given by

$$\binom{25}{11}\left(\frac{1}{5}\right)^{11}\left(\frac{4}{5}\right)^{14} + \cdots + \binom{25}{25}\left(\frac{1}{5}\right)^{25}\left(\frac{4}{5}\right)^{0}.$$

The calculation of this sum is prohibitively long; however, by the approximation method to be presented in Chapter 4, the probability can be shown to be approximately equal to .0035. The probability is *higher* under the assumption that p is greater than $\frac{1}{5}$; for example, it is very close to 1 when p is greater than $\frac{1}{2}$. If a subject consistently scores more than 10 correct guesses per 25, it seems unlikely that p is really $\frac{1}{5}$: it is more likely that p is greater than $\frac{1}{5}$, and hence, that the subject has ESP powers.

D. Industrial Acceptance Sampling

Suppose that a wholesale consumer orders a large lot of identical manufactured items from a producer; for example, a photographic supplies dealer orders a large lot of flashbulbs from a manufacturer. It is important for the producer and the consumer to know the quality of the bulbs in the lot: each bulb is either *defective*, that is, will not light properly, or, in the other case, is *nondefective*. We denote by p the fraction of defective items in the lot; this is unknown. On the one hand, some defective items are expected in a large lot because manufacturing processes are not perfect; on the other hand, the consumer does not want a lot with a large proportion of defectives. The only way in which he can be absolutely sure of the quality of the lot of bulbs he buys is to test all of them; however, since such complete testing would use up all the bulbs, he can test at most a very small fraction of them, and must compromise his certainty of the quality of the lot. The theory of the coin-tossing game has been used to evaluate procedures which enable the consumer to decide to accept or reject a lot on the evidence of the quality of a small sample of items from the lot.

Suppose that an item is selected from the lot in such a way that "no item is preferred over any other" in the process. We interpret this, in mathematical terms, to mean that the probability of drawing a defective is equal to the proportion of defectives in the lot. The actual condition of the item so drawn, defective or nondefective, is considered as the Outcome of the toss of a coin: H and T correspond to defective and nondefective, respectively, and the probability of H (defective) is taken to be equal to p. Suppose that a second item is selected from the lot after the first. If the lot is very large, the removal of the first barely changes the proportion of defectives in the lot; hence, we consider the condition of the second item as the Outcome of the toss of a coin with H and T corresponding to defective and nondefective, respectively, and with the *same probability p* of H (defective). Since the proportion of defectives in the lot is hardly changed by the outcome of the first draw, we say that the condition of the second item is practically uninfluenced by the condition of the first. The outcomes of the first two draws are assigned probabilities in accordance with the mathematical model (2.2). The successive drawing of items from the lot continues in this way until n have been drawn, where n is very small in comparison with the size of the lot. The outcomes of the n draws are assigned probabilities in the same way as the Outcomes in the coin-tossing game; hence, the probability distribution of the Number of H's (defectives) is the binomial distribution (2.5).

When the producer offers the consumer a lot of items, the latter is permitted to test a few and then to decide whether or not he wants to buy the lot. In the former case we say that he *accepts* the lot; in the latter he *rejects* it. Suppose that the consumer uses the following procedure to decide whether to accept or reject a lot: n bulbs are drawn "at random" from the lot and are tested; the lot is accepted or rejected accordingly as the number of defectives found is less than or equal to some predetermined number c, or greater than c, respectively. It follows that the probability that the lot is accepted is exactly the same as the probability that the Number of H's in n tosses of a coin is less than or equal to c: it is, by virtue of Definition 2.4,

$$(1 - p)^n + \binom{n}{1} p(1 - p)^{n-1} + \cdots + \binom{n}{c} p^c(1 - p)^{n-c}. \quad (2.7)$$

For each pair of numbers (n, c) formula (2.7) determines a certain probability depending on the value of p; this probability is denoted by $L(p)$. The procedure described here is called a sampling plan based on the pair of numbers (n, c); $L(p)$ represents the probability that the lot is accepted when p is the proportion of defectives.

Suppose that four items are sampled from the lot and that the plan calls for acceptance of the lot if not more than one is defective, and rejection if more than one is defective. The probability that the lot is accepted is, by formula (2.7), equal to

$$L(p) = (1 - p)^4 + 4p(1 - p)^3,$$

where p is the proportion of defectives in the lot. We find $L(p)$ for various values of p:

p	$L(p)$	p	$L(p)$
.0	1.0000	.6	.1792
.1	.9477	.7	.0837
.2	.8192	.8	.0272
.3	.6517	.9	.0037
.4	.4752	1.0	.0000
.5	.3125		

Note that the probability of acceptance decreases as the proportion of defectives increases from 0 to 1. There is a large probability of rejection (small probability of acceptance) when p is close to 1 and a small probability of rejection when p is small. The set of numbers $L(p)$, for values of p between 0 and 1 inclusive, is known as the *operating characteristic* of the sampling plan.

The particular choice of the numbers (n, c) is made by the producer and consumer; these numbers determine the operating characteristic. In the exercises at the end of this chapter the reader is asked to compute the operating characteristics for various sampling plans.

E. Preelection Polls

Suppose that the electorate of a population consists of members of two parties; let us refer to these as the Blue party and Green party, respectively. Let p be the proportion of Blue party members in the electorate. A poll taker selecting voters is like a quality engineer selecting items from a lot, as in the previous section: if we consider the Blues as H and the Greens as T, then the system of outcomes of the "random" selection of n voters, and the corresponding probabilities, are exactly the same as for n tosses of a coin. Here it is assumed, as in the case of a lot of items, that the electorate is very large and that the ratio of the number n to the number of voters is negligible. The number of Blues in a sample of n voters is like the Number of H's in n tosses of a coin; thus it has the binomial distribution (2.5).

2.2 EXERCISES

1. Find the probability of at least two girls in a family of five.

2. Which is more probable: at least one boy in a family of three, or at least two in a family of six?

3. A drug is known to cure 40 percent of patients. If five patients get the drug, what is the probability that at least half are cured? What if the drug is 80 percent effective?

4. If 60 percent of patients naturally recover from a disease, what is the probability that, in a group of seven patients, at least three will recover?

5. It is known that 50 percent of patients naturally recover from a disease. A drug is claimed to cure 90 percent of all patients. An experimental group of five patients gets the drug. What is the probability that at least four recover if

a) the drug is worthless and the true recovery rate is 50 percent; or
b) the drug is really 90 percent effective.

6. Under the assumption of "random guessing" ($p = \frac{1}{5}$), find the probability that a subject in the ESP experiment correctly guesses at least eight cards out of the first 10.

7. Five items are sampled from a lot ($n = 5$). The lot is rejected if more than one is defective ($c = 1$). Find the operating characteristic $L(p)$ for $p = .1, .2, .3, .4$.

8. Repeat Exercise 7 for $n = 5, c = 0$; then for $n = 6, c = 1$.

9. Seventy percent of an electorate belongs to the Blue party. If three voters are selected at random, what is the probability that at least two are Blue party members?

10. Sixty percent of an electorate belongs to the Blue party. Eight voters are selected at random. Find the probability that the number of Blue members is between four and seven, inclusive.

chapter 3

THE LAW OF LARGE NUMBERS
FOR COIN TOSSING

Now we shall establish the connection between the theoretical probability p attached to a coin and the actual outcomes of successive tosses. The main result of this chapter is the Law of Large Numbers: if p is the probability associated with a coin, and if the coin is tossed many times, then it is very likely that the ratio of the number of H's to the number of tosses will be very close to p. A precise formulation and proof of this principle require several ideas: the *expected value* and *variance* of the number of H's, and the *limit* of a sequence of real numbers. In Section 3.1 we define the first two, and show them to be equal to np and npq, respectively. In Section 3.2 we review the "limit" properties of the real number system. In Section 3.3 these results are used in the statement and proof of the Law; finally, applications are given.

3.1 EXPECTED VALUE AND VARIANCE OF THE NUMBER OF H's

Definition 3.1 The *expected value* of the Number of H's in n tosses of the coin is defined as the sum

$$1 \binom{n}{1} pq^{n-1} + 2 \binom{n}{2} p^2q^{n-2} + \cdots$$
$$+ k \binom{n}{k} p^kq^{n-k} + \cdots + n \binom{n}{n} p^nq^0; \qquad (3.1)$$

26

in words, it is a weighted average of the integers 1 through n, where each integer is given a weight equal to the corresponding term of the binomial distribution (formula 2.5). ⌡

In the case $n = 1$, the expected value of the Number of H's is p; when $n = 2$, it is

$$1 \binom{2}{1} pq + 2 \binom{2}{2} q^2.$$

The expected value of the Number of H's is called the "expected number of H's." It is the sum of the n products obtained by multiplying each possible number of H's $(0, 1, \ldots, n)$ by the corresponding probability; the product with the factor 0 contributes nothing to the sum. Although formula (3.1) is complicated, it turns out to have a simply expressed value, np.

Before formally proving this let us first present the proof for the particular case $n = 3$:

$$1 \binom{3}{1} pq^2 + 2 \binom{3}{2} p^2 q + 3 \binom{3}{3} p^3 = 3p.$$

First of all, $3p$ may be (artificially) written as $3p(p + q)^2$ because $p + q = 1$; then, the square is expanded; finally, each of its terms is multiplied by $3p$:

$$3p = 3p(p + q)^2 = 3pq^2 + 6p^2 q + 3p^3.$$

This is equal to the left-hand side of the former equation.

Proposition 3.1 *The expected number of H's, defined by formula (3.1), is equal to np.*

PROOF. The idea of the proof is: write np as $np(q + p)^{n-1}$, expand the binomial in accordance with the Binomial Formula, and then multiply the terms by np.

1. $np = np(q + p)^{n-1}$.

Reason. $(q + p)^{n-1} = 1^{n-1} = 1$ because $q + p = 1$.

2. $(q + p)^{n-1} = \binom{n-1}{0} q^{n-1} p^0 + \binom{n-1}{1} q^{n-2} p^1 + \cdots$

$$+ \binom{n-1}{k-1} q^{n-k} p^{k-1} + \cdots + \binom{n-1}{n-1} q^0 p^{n-1}.$$

Reason. Apply the Binomial Formula with q, p, $n - 1$, and $k - 1$ in place of A, B, N, and s, respectively.

3. $np \begin{pmatrix} n-1 \\ k-1 \end{pmatrix} q^{n-k}p^{k-1} = k \begin{pmatrix} n \\ k \end{pmatrix} p^k q^{n-k}.$

Reason. Apply formula (1.8), with n, k in place of N, s, respectively.

4. When each term in the binomial expansion of $(q + p)^{n-1}$ is multiplied by np, the resulting sum is the sum of the terms in formula (3.1).

Reason. Statement 3 shows that each *term* of the resulting sum is the corresponding *term* in formula (3.1).

5. The assertion of the proposition follows.

Reason. Combine Statements 1 and 4.   ◄

Here are examples based on Section 2.2.

Example 3.1 The expected value of the Number of boys in a family of five is the expected number of H's in five tosses of a balanced coin; here $n = 5$ and $p = \frac{1}{2}$, so that $np = 2.5$.   ◁

Example 3.2 A drug which is 35 percent effective is taken by 20 patients. The expected number of cured patients is $20(.35) = 7$.   ◁

Example 3.3 The expected number of correctly called cards in the ESP experiment is $25(\frac{1}{5})$ or 5, under the hypothesis of no ESP (random guessing).   ◁

Example 3.4 Ten items are selected from a large lot with two percent defectives; the expected number of defectives among the selected items is .2.   ◁

Example 3.5 A random sample of 100 voters is selected from a large electorate of which 53 percent belong to the Green party; the expected number of Green members in the voter sample is 53.   ◁

The concept of variance is now introduced. It is used in Section 3.3 as a device in the estimation of the sum of the probabilities in "tails" of the binomial distribution.

Definition 3.2 The variance of the Number of H's in n tosses of the coin is defined as the sum

$$(0 - np)^2 \begin{pmatrix} n \\ 0 \end{pmatrix} p^0 q^n + (1 - np)^2 \begin{pmatrix} n \\ 1 \end{pmatrix} p^1 q^{n-1} + \cdots$$
$$+ (k - np)^2 \begin{pmatrix} n \\ k \end{pmatrix} p^k q^{n-k} + \cdots + (n - np)^2 \begin{pmatrix} n \\ n \end{pmatrix} p^n q^0. \qquad (3.2)$$

This is a weighted average of squares of the differences between each of the integers $0, 1, \ldots, n$ and np, where each square is weighted by the corresponding term of the binomial distribution (formula 2.5).

In the case $n = 1$, the variance of the Number of H's is $(0 - p)^2 q + (1 - p)^2 p$; when $n = 2$, it is $(0 - 2p)^2 q^2 + (1 - 2p)^2 2pq + (2 - 2p)^2 p^2$.

The variance is the sum of the $n + 1$ products obtained by multiplying each squared difference by the probability of the corresponding number of H's; like the expected value, the variance has a simply expressed value, npq.

Before a formal proof, we compute the value of the sum (3.2) in the particular case $n = 3$; this calculation exhibits the important ideas of the formal proof. The sum in formula (3.2) is

$$(3p)^2 q^3 + (1 - 3p)^2 \cdot 3pq^2 + (2 - 3p)^2 \cdot 3p^2 q + (3 - 3p)^2 p^3.$$

Expand the squares and arrange the terms in three columns:

$1 \cdot 3pq^2$	$-2 \cdot 3p \cdot 3pq^2$	$(3p)^2 \cdot q^3$
$4 \cdot 3p^2 q$	$-2 \cdot 3p \cdot 6p^2 q$	$(3p)^2 \cdot 3pq^2$
$9p^3$	$-2 \cdot 3p \cdot 3p^3$	$(3p)^2 \cdot 3p^2 q$
		$(3p)^2 p^3$

Factor $(3p)^2$ from each term in the last column; then the sum of the remaining factors is $q^3 + 3pq^2 + 3p^2 q + p^3$; this is equal to $(q + p)^3 = 1$ by the Binomial Formula. It follows that the sum of the terms in the last column is $(3p)^2$. Now factor $-2 \cdot 3p$ from each term in the second column; then the sum of the remaining factors is $1 \cdot 3pq^2 + 2 \cdot 3p^2 q + 3p^2$, which, by Proposition 1 (for the case $n = 3$), is equal to $3p$. It follows that the sum of the terms in the second column is $-2 \cdot (3p)^2$. Now the sum of the terms in the first column may be written as $1^2 \cdot 3pq^2 + (2)^2 \cdot 3p^2 q + (3)^2 p^3$. For motives to be soon uncovered, we write $2^2 = 2 + 2(2 - 1)$, and $3^2 = 3 + 3(3 - 1)$. Inserting these in the sum, we get

$$1 \cdot 3pq + 2 \cdot 3p^2 q + 3p^3 + 2(2 - 1)3p^2 q + 3(3 - 1)p^3.$$

Here the sum of the first three terms is, by Proposition 1 (case $n = 3$), $3p$. The sum of the last two terms is $3 \cdot 2 \cdot p^2(q + p) = 3 \cdot 2 \cdot p^2$. The summary of the computation is that the sums of the terms in the three columns are $3 \cdot 2 \cdot p^2 + 3p$, $-2 \cdot (3p)^2$, and $(3p)^2$, respectively; hence, the grand sum is $3 \cdot 2p^2 + 3p - (3p)^2 = 3p[2p + 1 - 3p] = 3pq$.

Proposition 3.2 *The variance of the Number of H's, defined by formula (3.2), is equal to npq.*

omit
except
for special
cases n =1,2,3

PROOF. Consider the decomposition of the square $(k - np)^2$ into the sum of three terms:

$$k(k - 1) + k(1 - 2np) + n^2p^2;$$

this is verified by elementary algebraic operations. The sum in formula (3.2) defining the variance is the sum of all terms

$$(k - np)^2 \binom{n}{k} p^k q^{n-k}, \qquad k = 0, 1, \ldots, n.$$

By means of the above decomposition, the latter sum may be decomposed into a sum of *three* subsums:

subsum 1:   terms $k(k - 1) \binom{n}{k} p^k q^{n-k}$,

subsum 2:   terms $(1 - 2np)k \binom{n}{k} p^k q^{n-k}$,

subsum 3:   terms $n^2 p^2 \binom{n}{k} p^k q^{n-k}$,

$$k = 0, 1, \ldots, n.$$

1. Subsum 3 is equal to $n^2 p^2$.

Reason. When $n^2 p^2$ is factored from each term, the sum of the remaining factors is just the sum of the terms of the binomial distribution (formula 2.5), and is equal to 1.

2. Subsum 2 is equal to $(1 - 2np)np$.

Reason. When $(1 - 2np)$ is factored from each term, the sum of the remaining factors is indeed the sum in formula (3.1), which, by Proposition 3.1, is equal to np.

3. $n \binom{n-1}{k-1} = k \binom{n}{k}$.

Reason. Formula (1.8): n, k in place of N, s, respectively.

4. $(n - 1) \binom{n-2}{k-2} = (k - 1) \binom{n-1}{k-1}$.

Reason. This is the same as for the previous statement: $n - 1$, $k - 1$ replace n, k, respectively.

5. $n(n - 1) \binom{n-2}{k-2} = k(k - 1) \binom{n}{k}$.

Reason. Successively apply the previous two statements to the left-hand side of this equation:

$$n(n-1)\binom{n-2}{k-2} \rightarrow (k-1)n\binom{n-1}{k-1} \qquad \text{(Statement 5)}$$

$$(k-1)n\binom{n-1}{k-1} \rightarrow k(k-1)\binom{n}{k} \qquad \begin{array}{l}\text{(Statement 4; cf.}\\ \text{Section 1.2, Exercise 6)}\end{array}$$

6. $n(n-1)p^2 = n(n-1)p^2(q+p)^{n-2}$.

Reason. $q + p = 1$.

7. $(q+p)^{n-2}$ is equal to the sum of terms

$$\binom{n-2}{k-2} p^{k-2}q^{n-k}, \qquad k = 2, 3, \ldots, n.$$

Reason. Binomial Formula with $n-2$, $k-2$ in place of N, s, respectively.

8. Subsum 1 is equal to the sum of terms

$$n(n-1)\binom{n-2}{k-2} p^k q^{n-k}, \qquad k = 2, 3, \ldots, n.$$

Reason. Note that the terms of subsum 1 of indices $k = 0$ and $k = 1$ are both equal to 0; then apply Statement 5 to the remaining terms.

9. The sum in Statement 8 is equal to

$$n(n-1)p^2.$$

Reason. When $n(n-1)p^2$ is factored from each term, the sum of the remaining factors is that in Statement 7.

10. The sum defining the variance is equal to npq.

Reason. By virtue of Statements 8 and 9, subsum 1 is equal to $n(n-1)p^2$; and by Statements 1 and 2, respectively, subsums 3 and 2 are equal to n^2p^2 and $(1-2np)np$, respectively; thus, the grand sum is

$$n(n-1)p^2 + n^2p^2 + (1-2np)np = np - np^2 = npq. \qquad \blacktriangleleft$$

Example 3.6 The variance of the number of boys in a family of five children is $5(\frac{1}{2})(\frac{1}{2}) = \frac{5}{4}$ because $n = 5$, $p = q = \frac{1}{2}$.   ◁

Example 3.7 A drug which is 35 percent effective is taken by 20 patients. The variance of the number of cured patients is $(20)(.35)(.65) = 4.55$.   ◁

Example 3.8 The variance of the number of correctly called cards in the ESP experiment is $(25)(\frac{1}{5})(\frac{4}{5}) = 4$.   ◁

Example 3.9 Ten items are selected from a large lot with two percent defectives. The variance of the number of defectives among the selected items is $10(.02)(.98) = .196$.   ◁

Example 3.10 A random sample of 100 voters is selected from a large electorate of which 53 percent belong to the Green party. The variance of the number of Green members in the voter sample is $100(.53)(.47) = 24.91$. ◁

For each value of p (and q), the variance has a specified value. It happens that for each particular n, the variance npq is greatest when $p = q = \frac{1}{2}$, so that the variance never exceeds $n/4$:

> **Proposition 3.3** *The expression pq never exceeds $\frac{1}{4}$, where p and q are positive and have the sum 1.*

PROOF

1. $p^2 + q^2 = 1 - 2pq$.

Reason. $1 = (p + q)^2 = p^2 + 2pq + q^2$.

2. $2pq$ never exceeds $p^2 + q^2$.

Reason. The expression $(p - q)^2 = p^2 - 2pq + q^2$ is never less than 0.

3. $2pq$ never exceeds $1 - 2pq$.

Reason. Statements 1 and 2.

4. $4pq$ never exceeds 1.

Reason. Statement 3.   ◀

3.1 EXERCISES

1. Verify Propositions 3.1 and 3.2 in the special case $n = 4$.

2. Repeat Exercise 1 for the case $n = 5$.

3. Find the expected value and variance of the Number of H's in 80 tosses of a coin with $p = .6$.

siblings

4. Find the expected number of boys in a family of seven, and the variance.

5. A drug, 60 percent effective, is taken by 12 patients. Find the expected value and variance of the number of cured patients.

6. An electorate has 65 percent Blue party members. Find the expected value and variance of the number of Blue members in a sample of 150 voters.

7. Prove that the sum of the terms

$$(n - k) \binom{n}{k} p^k q^{n-k}$$

is *nq*.

8. By extending the method used in the evaluation of subsum 1 in the proof of Proposition 3.2, prove that the sum of the terms

$$k(k - 1)(k - 2) \binom{n}{k} p^k q^{n-k}, \qquad k = 3, 4, \ldots, n$$

is equal to $n(n - 1)(n - 2)p^3$.

9. Consider the following geometric interpretation of Proposition 3.3. Draw the graph of the function $y = x(1 - x)$ for values of x between 0 and 1. What is the highest point on the curve?

— *all of above covered on Test 1.* —

3.2 ELEMENTS OF THE THEORY OF LIMITS

The most interesting results of the theory of probability are those known as *limit theorems*. We shall illustrate such a result in the particular case of the coin-tossing game. If a fair coin is used, what is the probability that T never appears in the course of a game? If there are n tosses, then, in accordance with Definition 2.2, the probability of all H's is $(\frac{1}{2})^n$. If n is very large, then $(\frac{1}{2})^n$ is very small; furthermore, it gets progressively smaller as n gets larger; finally, $(\frac{1}{2})^n$ can be made arbitrarily small (close to 0) by the choice of a number n sufficiently large. We summarize this by saying that 0 *is the limit of* $(\frac{1}{2})^n$ *as n becomes infinite*; this can be also expressed as: $(\frac{1}{2})^n$ is approximately equal to 0 if n is very large. The limit theorem of probability that has just been proved is: the probability that T never appears in the course of a large number of tosses is approximately 0. This is a strictly mathematical version of the "law" that if a coin is tossed sufficiently many times, then T must eventually turn up.

Now we present the rudiments of the theory of limits. In the earlier statement that $(\frac{1}{2})^n$ gets "progressively smaller as n gets larger," the quantity $(\frac{1}{2})^n$ has been viewed not as an individual number but as part of a sequence of numbers $(\frac{1}{2})$, $(\frac{1}{2})^2$, $\ldots$, $(\frac{1}{2})^n$, $\ldots$. Our earlier statement means that this sequence has the property that each term of the sequence is at least as large as the succeeding term:

$$\left(\tfrac{1}{2}\right) \geq \left(\tfrac{1}{2}\right)^2 \geq \cdots \geq \left(\tfrac{1}{2}\right)^n \geq \cdots.$$

The other statement that $\left(\tfrac{1}{2}\right)^n$ can be made arbitrarily small by the choice of n sufficiently large has the following precise meaning: for any (arbitrarily small) positive number ϵ, there is an index n (sufficiently large) such that the corresponding member of the sequence $\left(\tfrac{1}{2}\right)^n$ is less than ϵ. This can be affirmed as follows: for any positive number ϵ, any term $\left(\tfrac{1}{2}\right)^n$ of the sequence whose index n is greater than $\log(1/\epsilon)/\log 2$ is less than ϵ; in fact, the inequality

$$n > \log(1/\epsilon)/\log 2,$$

which is the same as $n \log 2 > \log(1/\epsilon)$, is equivalent to

$$n \log\left(\tfrac{1}{2}\right) < \log \epsilon,$$

or to

$$\left(\tfrac{1}{2}\right)^n < \epsilon.$$

This particular case is extended to the following general definition. (The "numbers" in the next three definitions are real numbers. The axioms of the real number system are reviewed below.)

Definition 3.3 Let $x_1, x_2, \ldots, x_n, \ldots$ be a sequence of nonnegative numbers such that each is at least as great as its successor:

$$x_1 \geq x_2 \geq \cdots \geq x_n \geq \cdots.$$

We say that the *sequence has the limit* 0 if for any (arbitrarily small) positive number ϵ, there is an integer n (large enough) such that the corresponding term x_n is less than ϵ.

Example 3.11 Put $x_1 = 1$, $x_2 = \tfrac{1}{2}, \ldots, x_n = 1/n, \ldots$; then the sequence has the limit 0; indeed, for any number $\epsilon > 0$, an integer n greater than $1/\epsilon$ satisfies the requirement of Definition 3.3: $x_n = (1/n) < \epsilon$ if $n > (1/\epsilon)$.

Another illustration is furnished by the sequence $x_1 = 1$, $x_2 = \tfrac{1}{4}, \ldots,$ $x_n = 1/n^2, \ldots$: for any positive number ϵ, an integer n greater than $1/\sqrt{\epsilon}$ has the property that $x_n = 1/n^2 < \epsilon$; thus, this sequence has the limit 0. ◁

Definition 3.3 was restricted to sequences of nonnegative numbers having the property that each term is at least as large as its successor in the sequence. Now we extend this definition to an *arbitrary* sequence of nonnegative numbers. The point of the following definition is that a sequence is said to have the limit 0 if it is "smaller" than some other sequence having the limit 0 under Definition 3.3.

> **Definition 3.4** Let $y_1, y_2, \ldots, y_n, \ldots$ be a sequence of nonnegative numbers. If there exists (possibly another) sequence $x_1, x_2, \ldots, x_n, \ldots$ which
>
> a) satisfies the hypothesis of Definition 3.3 and has the limit 0 in the sense of that definition; and
> b) each term of the y-sequence is less than or equal to the corresponding term of the x-sequence: $y_1 \leq x_1, y_2 \leq x_2, \ldots, y_n \leq x_n, \ldots,$
>
> then the y-sequence is said to have the limit 0.

Applications of this definition appear in the next section; however, we now present a synthetic example for illustration.

Example 3.12 Consider the two sequences given in Example 3.11:

$$1, \tfrac{1}{2}, \ldots, 1/n, \ldots; \qquad 1, \tfrac{1}{4}, \ldots, 1/n^2, \ldots.$$

Combine these in a single sequence in this way: choose alternate elements from the two sequences: $1, \tfrac{1}{4}, \tfrac{1}{3}, \tfrac{1}{16}, \ldots, 1/n, 1/(n+1)^2, \ldots.$ This sequence does not satisfy the hypothesis of Definition 3.3 because the odd-numbered terms are larger than their immediate predecessors in the combined sequence; however, it *does* fulfill the requirements of Definition 3.4 because each term of the combined sequence does not exceed the corresponding term of the first sequence: $1 \leq 1, \tfrac{1}{4} \leq \tfrac{1}{2}, \tfrac{1}{3} \leq \tfrac{1}{3}, \tfrac{1}{16} \leq \tfrac{1}{4}, \ldots.$ ◁

The previous two definitions defined only the limit 0, and only for sequences of nonnegative numbers; the following defines limit in a more general way.

> **Definition 3.5** Let $z_1, z_2, \ldots, z_n, \ldots$ be an arbitrary sequence of numbers, and L a number. The sequence is said to have the limit L if the sequence of absolute differences $y_1 = |z_1 - L|, y_2 = |z_2 - L|, \ldots,$ $y_n = |z_n - L|, \ldots$ (which is a sequence of nonnegative numbers) has a limit in the sense of Definition 3.4.

Example 3.13 Put $z_1 = 1, z_2 = 1 + \tfrac{1}{2}, z_3 = 1 - \tfrac{1}{3}, z_4 = 1 + \tfrac{1}{4}, \ldots,$ $z_n = 1 + [(-1)^n/n], \ldots.$ This sequence has the limit $L = 1$ because the sequence of absolute differences $y_1 = 1, y_2 = \tfrac{1}{2}, y_3 = \tfrac{1}{3}, \ldots, y_n = 1/n$ has the limit 0 under either Definition 3.3 or 3.4. ◁

Note that the three definitions of a limit are in increasing order of generality. A sequence having the limit 0 in the sense of Definition 3.3 also has the limit 0 in the sense of Definition 3.4; and one having the limit 0 in the sense of Definition 3.4 also has it in the sense of Definition 3.5. (See Exercise 3 below.)

The discussion of limits just given is sufficient for the formulation and proof of all of the forthcoming limit theorems—with one exception: Proposition 6.5 below, which is concerned with the solution of the gambler's ruin problem. In order to complete the exposition, we recall the axiomatic description of the real number system; a more detailed sketch may be obtained from a standard text on calculus. Our purpose in citing these axioms is to show how they may be used to prove the *existence* of the limit of a certain sequence without actually *computing* the limit.

The real number system is the smallest set of numbers having the following four properties:

1) It contains the set of rational numbers.

2) It has the algebraic properties of the rational numbers: closure under the four rational binary operations of addition, subtraction, multiplication, and division; and satisfies the commutative, associative, and distributive laws.

3) It has the *order* properties of the rational numbers: there is an order relation $<$ ("less than") on the real numbers such that for each pair x, y either $x < y$ or $y < x$ or $x = y$.

4) The fourth property, which distinguishes the set of real numbers from the set of rational numbers, is called the *completeness* property. A set B of real numbers is said to have the real number b as an *upper bound* (or b is said to be an upper bound for B) if every element of B is not greater than b. According to this definition, an upper bound b does not necessarily belong to B; furthermore, a set may have more than one upper bound (and usually does) if it has one. A number c is called a *least upper bound* for the set B if (a) c is an upper bound and (b) no other upper bound is smaller than c. The completeness property of the real number system is: Every set of real numbers having an upper bound necessarily has a least upper bound (which is unique).

Here is a simple application of the completeness property: the definition of the square root of 2. It is not a rational number, but is a real number defined as the least upper bound of the set of all rational numbers whose squares are less than 2. The existence of this least upper bound is a consequence of property 4; indeed, the set of all rational numbers whose squares are less than 2 has an upper bound (e.g., 3), so that it has a *least* upper bound.

The concluding result of this section, which follows from the completeness property and which will be applied in Chapter 6, is portrayed in the following example. A man, starting at a given distance from a wall, takes an infinite number of steps toward the wall without reaching it in a finite number of steps; in other words, each step carries him less than the remaining distance to the wall. Our principle is: either he gets arbitrarily close to the wall in sufficiently many steps (the wall is the "limit" of his sequence of steps) or else there is a point *before* the wall to which he comes arbitrarily close in sufficiently many steps. The mathematical version of this conclusion is:

If $x_1, x_2, \ldots$ is a sequence of real numbers and b a real number such that (a) each successive term of the sequence is at least equal to its predecessor and (b) no term in the sequence exceeds b, then there exists a real number L which is the limit of the sequence.

The proof is outlined in Exercise 7 below; applications are contained in several other exercises.

3.2 EXERCISES

1. Find the limit L of each sequence, stating which definition has been employed:

a) $x_1 = (\frac{1}{2}) + 3, x_2 = (\frac{1}{2})^2 + 3, \ldots, x_n = (\frac{1}{2})^n + 3, \ldots$.

b) $x_1 = (\frac{1}{2}) - 3, x_2 = (\frac{1}{2})^2 - 3, \ldots, x_n = (\frac{1}{2})^n - 3, \ldots$.

c) $x_1 = 4 - (.1), x_2 = 4 + (.1)^2, \ldots, x_n = 4 + (-.1)^n, \ldots$.

d) $x_1 = 1, x_2 = -\frac{1}{3}, \ldots, x_n = (-\frac{1}{3})^n, \ldots$.

e) $x_1 = 2, x_2 = 2, \ldots, x_n = 2, \ldots$.

2. Show that the "infinite" decimal expansion .333 ... (infinitely many digits 3) represents the number $\frac{1}{3}$ in the following sense: the limit of the sequence of approximating decimals .3, .33, .333, etc., is $\frac{1}{3}$.

3. Prove: If a sequence of real numbers has the limit 0 in the sense of Definition 3.3, then it also has it in the sense of Definition 3.4. Show that the corresponding statement for Definitions 3.4 and 3.5, respectively, is also true.

4. In algebra it is shown that the value of the infinite geometric series

$$1 + r + r^2 + \cdots$$

is equal to $1/(1 - r)$ if $-1 < r < 1$. Prove this result in terms of limits: show that the sequence of partial sums

$$1, 1 + r, 1 + r + r^2, \ldots, 1 + r + \cdots + r^n, \ldots$$

has the limit $1/(1 - r)$. In the proof use the formula

$$(1 - r)(1 + r + \cdots + r^n) = 1 - r^{n+1}.$$

5. Show that the repeating infinite decimal below has the indicated value:

$$.378378378 \ldots = 378/999.$$

[Hint: Write the decimal as an infinite geometric series: $.378(1 + r + r^2 + \cdots)$, where $r = 10^{-3}$, and then apply the result of the previous exercise.]

6. Prove: The sequence of partial decimal expansions

$$.a_1, .a_1a_2, .a_1a_2a_3, \ldots, .a_1 \ldots a_n, \ldots$$

obtained from an infinite decimal $.a_1a_2 \ldots$ has a limit. [Show that the successive terms of the sequence are at least as large as their respective predecessors; and that each term of the sequence is not greater than $(a_1 + 1)/10$; and then use the consequence of the completeness property.]

7. Complete the details of this sketch of the proof of the consequence of the completeness property; here $x_1, x_2, \ldots$ is a sequence of real numbers and b a real number such that each successive term of the sequence is at least as large as its predecessor, and none exceeds b.

a) The set of numbers $x_1, x_2, \ldots$ has a least upper bound.
b) Let L be the least upper bound; then for every number $\epsilon > 0$, there is an index n sufficiently large that the corresponding difference $L - x_n$ is less than ϵ. (If not, L would not be a least upper bound.)
c) L is the limit of the sequence.

8. Let $p_1, p_2, \ldots$ be a sequence of probabilities such that each successive one is at least as large as its predecessor. Why does the sequence necessarily have a limit?

3.3 STATEMENT AND PROOF OF THE LAW OF LARGE NUMBERS FOR COIN TOSSING; APPLICATIONS

Now we discuss one of the earliest important limit theorems of the theory of probability: the law of large numbers for coin tossing. It was discovered by James Bernoulli (1654–1705). The proof he originally gave for this law was later simplified by the use of a theorem of P. L. Chebychev (1821–1894). The following proposition is a particular case of the "Chebychev inequality":

Proposition 3.4 Let d be an arbitrary positive real number. The probability that the Number of H's in n tosses of the coin is between the two numbers $np - nd$ and $np + nd$ is at least as large as

$$1 - (pq/nd^2). \qquad (3.3)$$

(Reread Definition 2.4 before going on.)

This proposition has immediate applications but a complicated proof; hence, we postpone the latter for a while.

Example 3.14 A balanced coin ($p = q = \frac{1}{2}$) is tossed 100 times; choose (arbitrarily) $d = .1$. The quantities $np - nd$, $np + nd$, and $1 - (pq/nd^2)$ mentioned in the proposition are equal to

$$100(\tfrac{1}{2}) - 100(.1) = 40,$$
$$100(\tfrac{1}{2}) + 100(.1) = 60,$$
$$1 - [\tfrac{1}{2} \cdot \tfrac{1}{2}/100(.1)^2] = \tfrac{3}{4},$$

respectively. The proposition implies: the probability that the Number of H's is between 40 and 60 is at least $\frac{3}{4}$. Another way of stating this is: the probability is at least $\frac{3}{4}$ that the ratio

Number of H's in 100 tosses/100

is between .4 and .6. ◁

Example 3.15 A coin with $p = .3$ is tossed 100 times; choose $d = .1$. The quantities mentioned in the proposition have the values

$$np - nd = 100(.3 - .1) = 20,$$
$$np + nd = 100(.3 + .1) = 40,$$
$$1 - (pq/nd^2) = 1 - [.3(.7)/100(.1)^2] = .79,$$

respectively. The proposition implies that the probability that the Number of H's is between 20 and 40 is at least .79; or, in other words, the probability is at least .79 that the *ratio* of the Number of H's in 100 tosses to the number (100) of tosses is between .2 and .4. ◁

Example 3.16 Let d be decreased to .01, and n increased to 100,000; and reconsider Example 3.14, $p = \frac{1}{2}$. The quantities in the proposition are now

$$np - nd = 100,000(\tfrac{1}{2} - .01) = 49,000,$$
$$np + nd = 100,000(\tfrac{1}{2} + .01) = 51,000,$$
$$1 - (pq/nd^2) = 1 - [1/4(100,000)(.01)^2] = .975,$$

respectively. The probability is at least .975 that the ratio

Number of H's in 100,000 tosses/100,000

is between .49 and .51. If d is kept as .01 and n increased to 1,000,000, the probability is at least .9975 that the ratio of the Number of H's to the number of tosses is between .49 and .51. ◁

In Example 3.15, when d is reduced to .01 and n is increased to 100,000, the probability is at least .979 that the ratio of the Number of H's to the number of tosses is between .29 and .31. When n is increased to 1,000,000, the corresponding probability is at least .9979.

These examples indicate that as n gets larger and larger, the probability that the average number of H's (ratio of Number of H's to number of tosses) is "close" to the theoretical probability p is "nearly" 1. This fact can be precisely stated in the form of a limit theorem:

Law of Large Numbers. *For a fixed positive number d, let P_n be the probability that the ratio of the Number of H's to the number of tosses is between $p - d$ and $p + d$, $n = 1, 2, \ldots$; then the limit of the sequence $P_1, P_2, \ldots, P_n, \ldots$ is 1.*

Before giving the proof, we wish to refer the reader to the definition of limit in the previous section, and, in particular, to Definition 3.5. In order to prove the Law of Large Numbers, we shall show that the sequence of absolute differences $|P_1 - 1|, |P_2 - 1|, \ldots, |P_n - 1|, \ldots$ has the limit 0 under Definition 3.4; more specifically, we shall show that the sequence of absolute differences is "smaller" than the sequence $pq/d^2, pq/2d^2, \ldots,$ $pq/nd^2, \ldots$, which has the limit 0 under Definition 3.3. The proof is based on Proposition 3.4, which will be proved further on.

PROOF OF THE LAW OF LARGE NUMBERS

1. The probability P_n that *the ratio of the Number of H's in n tosses to n is between $p - d$ and $p + d$* is exactly the same as the probability that *the Number of H's is between $np - nd$ and $np + nd$.*

Reason. This is a simple algebraic consequence of the definition of "ratio."

2. The absolute difference $|1 - P_n|$ is less than or equal to pq/nd^2, $n = 1, 2, \ldots$.

Reason. On one hand, P_n is not greater than 1 because it is a probability; on the other hand, P_n is at least equal to expression (3.3): $1 - (pq/nd^2)$.

3. The sequence of absolute difference $|1 - P_1|, |1 - P_2|, \ldots, |1 - P_n|, \ldots$ has the limit 0.

Reason. This follows from the fact that the sequence is "smaller" than the sequence $pq/d^2, \ldots, pq/2d^2, \ldots, pq/nd^2, \ldots$ by virtue of Statement 3.

4. The sequence $P_1, P_2, \ldots, P_n, \ldots$ has the limit 1.

Reason. Statement 3 and Definition 3.5. ◀

Before starting the formal proof of Proposition 3.4, upon which the Law of Large Numbers is based, we illustrate a general algebraic principle by means of a simple example. Let $a_1, a_2, a_3, b_1, b_2, b_3$, and c be arbitrary nonnegative numbers; our only assumptions about them are:

$$c \leq a_1, \quad c \leq a_2, \quad a_3 \leq c.$$

We infer that the *sum $a_1b_1 + a_2b_2 + a_3b_3$ is at least as large as $c(b_1 + b_2)$*. This follows from the facts that the former sum is at least as large as $a_1b_1 + a_2b_2$, and that $a_1b_1 + a_2b_2$ is at least as large as $cb_1 + cb_2$. Now we state and prove the general principle.

> **Proposition 3.5** Let $(a_0, a_1, \ldots, a_n)$ and $(b_0, b_1, \ldots, b_n)$ be two sets of nonnegative numbers, and c a positive number; and form the sum $a_0b_0 + a_1b_1 + \cdots + a_nb_n$. Denote by B the sum of the factors $b_0, b_1, \ldots$ whose corresponding coefficients $a_0, a_1, \ldots$ are at least as large as c. We conclude that the sum $a_0b_0 + \cdots + a_nb_n$ is at least as large as cB.

PROOF

1. The sum of only those terms of the sum $a_0b_0 + \cdots + a_nb_n$ in which the a-factors are at least as large as c is itself at least as large as c multiplied by the sum of the corresponding b-factors, that is, cB.

Reason. If the numbers $a_0, a_1, \ldots$ are at least equal to c, then the sum $a_0b_0 + a_1b_1 + \cdots$ is at least equal to $cb_0 + cb_1 + \cdots$ because the b-factors are nonnegative.

2. The sum of *all* the terms $a_0b_0 + \cdots + a_nb_n$ is at least equal to the sum of only those terms whose a-factors are at least equal to c.

Reason. A sum of nonnegative terms cannot be increased as the result of the removal of terms.

3. The sum $a_0b_0 + \cdots + a_nb_n$ is at least as large as cB.

Reason. Statements 1 and 2. ◀

We now use Proposition 3.5 to prove Proposition 3.4: the fundamental idea of the proof is the application of the former proposition to the sum (3.2) defining the variance. The factors $(k - np)^2$ are taken as the "a-factors" and $\binom{n}{k}p^kq^{n-k}$ as the "b-factors"; n^2d^2 is taken to be c.

PROOF OF PROPOSITION 3.4

1. The sum of the terms $\binom{n}{k}p^k q^{n-k}$, $k = 0, 1, \ldots, n$ in which the index k satisfies the inequality $(k - np)^2 \geq n^2 d^2$ is equal to 1 minus the sum of the terms for which the alternative inequality $(k - np)^2 < n^2 d^2$ holds.

Reason. The sum of the terms of the binomial distribution is 1, as shown in Chapter 2.

2. The sum (3.2), which is of terms $(k - np)^2 \binom{n}{k}p^k q^{n-k}$, is greater than or equal to $n^2 d^2$ multiplied by the first sum in Statement 1.

Reason. Apply Proposition 3.5 with $c = n^2 d^2$, and with $(k - np)^2$ and $\binom{n}{k}p^k q^{n-k}$ as a-factors and b-factors, respectively.

3. The inequality $(k - np)^2 < n^2 d^2$ is satisfied if and only if k is between $np - nd$ and $np + nd$.

Reason. The quantity $(k - np)^2$ is less than $n^2 d^2$ if and only if the same relation exists for their corresponding positive square roots: $|k - np| < nd$. The latter is equivalent to the double inequality $np - nd < k < np + nd$.

4. The sum of those terms of the binomial distribution for which the index k is between $np - nd$ and $np + nd$ is equal to the probability that the Number of H's is between $np - nd$ and $np + nd$.

Reason. Definition 2.4.

5. The sum (3.2) is greater than or equal to $n^2 d^2$ multiplied by 1 minus the probability that the Number of H's is between $np - nd$ and $np + nd$.

Reason. Statements 2, 3, and 4.

6. The quantity npq is greater than or equal to $n^2 d^2$ multiplied by 1 minus the probability that the Number of H's is between $np - nd$ and $np + nd$.

Reason. Statement 5 and Proposition 3.2.

7. The previous statement completes the proof.

Reason. Let P_n stand for the probability described in Statement 6: the latter states that $npq \geq n^2 d^2 (1 - P_n)$, which is also equivalent to the conclusion of Proposition 3.4, namely, $P_n \geq 1 - (pq/nd^2)$.   ◀

The Law of Large Numbers ties the probability p that the coin turns up H on a single toss to the long-run frequency of H in a large number of tosses. This law is a mathematical idealization of the popular "law of averages"; a typical application of the latter is the "law" that if a fair coin is tossed many times, then the ratio of the Number of H's to the number of tosses "will be one-half." But it has a much more profound application: if a coin with an *unknown* probability p is tossed many times, then p can be *estimated* by the ratio of the Number of H's to the number of tosses; furthermore, Proposition 3.4 can be used to measure the "accuracy" of the estimate.

Example 3.17 A poll taker wants to estimate the proportion p of Blue party members in the electorate for the purpose of predicting the outcome of the election (Section 2.2E). Each voter selected at random from the population is like the Outcome of the toss of a coin with probability p of H. Suppose that the poll taker wants the margin of error in his estimate of p to be no larger than .01; that is, he is willing to be wrong by no more than one percentage point. Put $d = .01$ in Proposition 3.4: the probability that the Number of Blue party members in a sample of n voters is between $n(p - .01)$ and $n(p + .01)$ is at least as large as expression (3.3) with $d = .01$; in other words, the probability that the *proportion of Blue members in the sample* differs from the proportion p in the electorate by at most .01 is at least equal to

$$1 - pq/n(.01)^2.$$

This quantity is *at least* equal to

$$1 - 1/4n(.01)^2; \tag{3.4}$$

indeed, by Proposition 3.2, pq does not exceed $\frac{1}{4}$. Suppose a sample of $n = 25,000$ voters is taken: in accordance with formula (3.4), the probability is at least $1 - 1/4(25,000)(.0001) = .9$ that the proportion of Blue members in the sample differs from the proportion p in the electorate by not more than $d = .01$. If the sample is increased to $n = 250,000$, the corresponding probability is at least .99. It is evident from expression (3.4) that the probability that the proportion in the sample differs from p by at most .01 can be pushed as close to 1 as desired by the choice of a sufficiently large sample size n; or, in mathematical terms, the *limit of the probability is* 1. ◁

The numerical results obtainable from Proposition 3.4 can be significantly improved by the use of the *normal approximation* theory in the next chapter. The sample size demanded of the poll taker in the previous example is much too large to be used in practice; however, we shall see that such large samples are theoretically unnecessary and that much smaller samples of practical size give the same error margins and probability bounds. For this reason, Proposition 3.4 is of theoretical and pedagogical value rather than of immediate practical value: it provides the steps to an understanding of the Law of Large Numbers.

3.3 EXERCISES

1. A coin with probability $p = .2$ is tossed 10,000 times. Use Proposition 3.4 to find the minimum probability that the Number of H's is between 1500 and 2500. (Hint: Choose $d = .05$.)

2. A coin with probability $p = .6$ is tossed 100,000 times. Find the minimum probability that the Number of H's is between 59,000 and 61,000.

3. In one million births, how large is the probability that the fraction of males is between .495 and .505?

4. Repeat Exercise 3 for ten million births.

5. If a subject in the ESP experiment (Section 2.2C) does *not* have powers of ESP, how large is the probability that he correctly guesses from 100 to 300 cards out of a total of 1000?

6. A polltaker wants to estimate the proportion p of Blue party members in the electorate. He wants his margin of error d to be at most equal to .02 and his probability of being correct to be at least equal to .99. How large a sample of voters does he need? (Hint: Follow Example 3.17 with $d = .02$. Solve the equation $1 - 1/4n(.02)^2 = .99$ for the unknown n.)

7. Repeat Exercise 6 with a margin of error $d = .03$ and a minimum probability .97 of being correct.

8. Is Proposition 3.5 still true if any of the a-terms or b-terms are negative?

9. Using the same methods as for Proposition 3.4, prove the following generalization: For any even integer $m > 1$ the probability that the number of H's in n tosses of the coin is between $np - nd$ and $np + nd$ is at least as large as

$$1 - \left[\text{sum of terms: } (k - np)^m \binom{n}{k} p^k q^{n-k} \right] / n^m d^m.$$

(When $m = 2$, this reduces to Proposition 3.4.)

10. Is the result of Exercise 9 still true for odd integers m? (Hint: Cf. Exercise 8.)

THE NORMAL APPROXIMATION
TO THE BINOMIAL DISTRIBUTION

Let a and b be two fixed positive numbers. If a coin is tossed n times, then the probability that the Number of H's is between a and b inclusive is, by Definition 2.4, the sum of the terms of the binomial distribution (formula 2.5), in which the index k is between a and b inclusive. If n is very large, then the computation of this probability is a long process; therefore, an approximation is desirable and, fortunately, available.

We introduce the *standard normal curve:* it is the graph of the function

$$y = (1/\sqrt{2\pi})e^{-(x^2/2)}, \tag{4.1}$$

where e is an irrational number whose numerical value is approximately 2.72; in other words, for each number x, the height of the curve in the xy-plane is given by the value of y in Eq. (4.1). Figure 4.1 is a pictorial representation of the curve. As visible from the picture, the curve has a maximum at $x = 0$, is symmetric about that point, is "bell-shaped," and *seems* to vanish for values of x either to the left of -3 or to the right of $+3$. In reality, it never vanishes but gets progressively closer to the x-axis as x moves indefinitely out in the positive and negative directions. It can be

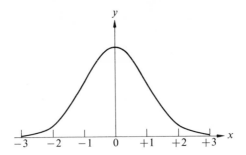

Fig. 4.1. The standard normal curve.

shown by methods of integral calculus that the total area under the curve is 1 square unit. The areas under portions of the curve will be used to approximate sums of probabilities in the binomial distribution. Since these areas are so important in applications, we now describe the area characteristics of the curve and methods of computing areas from a table.

The area under the curve between the values $x = 0$ and $x = +1$ is approximately .34; therefore, by the symmetry of the curve, it is also the area under the curve from $x = -1$ to $x = 0$; hence, the area from $x = -1$ to $x = +1$ is about .68. The areas between $x = +1$ and $x = +2$ and between $x = -2$ and $x = -1$ are both approximately .14; therefore, the total area from $x = -2$ to $x = +2$ is approximately .96. The total area from $x = -3$ to $x = +3$ is more than .99.

Table II (p. 202) indicates area under the curve from $x = 0$ to $x = z$ for positive numbers z; using these, one can compute the area under the curve between any two values. Here are some of the rules used for such computations:

1. The area under the curve to the right (or to the left) of the point $x = 0$ is $\frac{1}{2}$ because the curve is symmetric about 0 and the total area is 1.

2. The area to the right of a positive number z is the difference between the area to the right of $x = 0$ (that is, $\frac{1}{2}$) and the area from $x = 0$ to $x = z$, which is given in the table.

3. The area to the left of a negative number $x = -z$ is equal, by the symmetry of the curve, to the area to the right of the positive number $x = +z$; the latter case has been considered in No. 2 above.

4. The area to the left of a positive number $x = z$ is equal to $\frac{1}{2}$ plus the area from 0 to z.

5. The area to the right of a negative number $x = -z$ is equal, by symmetry, to the area to the left of the positive number $x = +z$; the latter case has been considered in No. 4 above. ⌡

To find the area under the curve between any two numbers A and B, we use the additive property of area measurement that the area under the curve over two nonoverlapping intervals is equal to the sum of the areas over the two intervals. •

6. *A and B both positive, $A < B$.* The area between A and B is the difference of the areas between 0 and B and between 0 and A. The latter are obtained from the table.

7. *$-A$ is negative and B is positive.* The area from $-A$ to B is equal to the sum of the areas from 0 to $+A$ and from 0 to B; these are obtained from the table.

8. *$-A$ and $-B$ are negative, $-B < -A$.* By symmetry of the curve, the area between $-B$ and $-A$ is equal to the area between $+A$ and $+B$, already considered in No. 6.

These rules will be illustrated in the numerical examples that will be given. We now state the normal approximation to the binomial distribution, which was proved in successive stages by A. deMoivre (1667–1754) and P. S. Laplace (1749–1827).

> **Normal Approximation Theorem.** *The probability that the Number of H's is between the numbers* inclusive
>
> $$a = np + A\sqrt{npq} + \tfrac{1}{2} \quad \text{and} \quad b = np + B\sqrt{npq} - \tfrac{1}{2} \quad (4.2)$$
>
> *is approximately equal to the area under the standard normal curve between $x = A$ and $x = B$ if n is large; more precisely, the sequence of probabilities has the limit equal to that area.*⌋

The content of this theorem is that the sum of the terms of the binomial distribution whose indices k are between $np + A\sqrt{npq} + \tfrac{1}{2}$ and $np + B\sqrt{npq} - \tfrac{1}{2}$ is approximately equal to the appropriate area under the standard normal curve. The proof of the validity of this approximation is based on calculus, which is beyond the prerequisites of this book (cf. W. Feller, *An Introduction to Probability Theory and Its Applications*, 3rd ed., Vol. 1, New York: Wiley, 1968, Chapter VII); however, in a later chapter, we offer an "experimental" method of confirming the normal approximation using the Law of Large Numbers. The approximation will be used only for the numerical illustrations in this chapter and Chapter 13, and in the discussion of the normal distribution in Chapter 10.

Let us illustrate this theorem in the special case where $n = 10,000$, $p = q = \frac{1}{2}$, $a = 4950$, and $b = 5100$: we seek the (approximate) probability that the Number of H's in 10,000 tosses of a balanced coin is between 4950 and 5100.

Step 1. Solve Eqs. (4.2) for A and B:

$$A = \frac{a - np - \frac{1}{2}}{(npq)^{1/2}} = \frac{4950 - 5000 - .5}{[10,000(\frac{1}{2})(\frac{1}{2})]^{1/2}} = -1.1,$$

$$B = \frac{b - np + \frac{1}{2}}{(npq)^{1/2}} = \frac{5100 - 5000 + .5}{[10,000(\frac{1}{2})(\frac{1}{2})]^{1/2}} = +2.1.$$

Step 2. Find the area under the standard normal curve from $A = -1.1$ to $B = 2.1$. Use rule 7 above for the calculation: it is the sum of

a) the area from 0 to $x = 1.1$, which, from Table II, is .3643; and
b) the area from 0 to $x = 2.1$, which, from Table II, is .4821.

It follows that the probability we seek is equal to $.3643 + .4821 = .8464$.

Here are some illustrations of the use of the normal approximation in the applications of the coin-tossing game in Chapter 2.

Example 4.1 *Sex of a newborn child.* Suppose that 100 babies are delivered every week in a certain hospital. What is the probability that more than 58 of these are boys? This is the same as the probability that more than 58 H's appear in 100 tosses of a balanced coin. We apply the normal approximation to find the probability that the Number of H's is between $a = 59$ and $b = 100$.

Step 1. Solve Eqs. (4.2) for A and B:

$$A = \frac{59 - 50 - .5}{[100(\frac{1}{2})(\frac{1}{2})]^{1/2}} = 1.7,$$

$$B = \frac{100 - 50 + .5}{[100(\frac{1}{2})(\frac{1}{2})]^{1/2}} = 10.1.$$

Step 2. Find the area under the standard normal curve from $A = 1.7$ to $B = 10.1$. Since the area to the right of $x = 10.1$ is less than .0001, the area from 1.7 to 10.1 is, to four decimal places, equal to the area to the right of 1.7. According to rule 2, this is the difference between $\frac{1}{2}$ and the area from $x = 0$ to $x = 1.7$, which, from the table, is .4554; hence, the area to the right of $x = 1.7$ is .0446. This is the approximate probability that we were seeking. ◁

Example 4.2 *Medical testing.* Suppose that 40 percent of patients naturally recover from a disease. A medical research worker wants to test the effectiveness of a drug proposed as a treatment: the sponsor claims that it will increase the rate of recovery to 80 percent. Sixty patients are treated with the drug. The investigator decides that he will accept the claim of the sponsor if at least 35 of the 60 patients are cured by the drug; on the other hand, he will reject it as worthless if fewer than 35 recover. For the analysis of this problem let us refer to the coin-tossing model in Chapter 2. Under the hypothesis that the drug is 80 percent effective, the actual effect on each patient is like the Outcome of the toss of a coin with $p = .8$; on the other hand, under the hypothesis that the drug is worthless, the coin has probability $p = .4$. We shall compute the following two probabilities:

a) the probability that at least 35 patients recover even though the drug is really worthless; this is the probability that the claim of the sponsor is *mistakenly accepted*; and

b) the probability that fewer than 35 patients recover even though the drug is really 80 percent effective; this is the probability that the claim of the sponsor is *mistakenly rejected.*

The first probability is the probability of at least 35 H's in 60 tosses of a coin with $p = .8$; to calculate it we use the normal approximation with $n = 60$, $p = .8$, $a = 35$, $b = 60$.

Step 1. Find A and B in Eqs. (4.2):

$$A = \frac{35 - (60)(.4) - .5}{[60(.4)(.6)]^{1/2}} = \frac{10.5}{3.8} = 2.8 \quad \text{(approximately)},$$

$$B = \frac{60 - 60(.4) + .5}{[60(.4)(.6)]^{1/2}} = \frac{36.5}{3.8} = 9.6 \quad \text{(approximately)}.$$

Step 2. The area under the standard normal curve between $A = 2.8$ and $B = 9.6$ is, to four decimal places, the same as the area to the right of $A = 2.8$; by rule 2, it is the difference between $\frac{1}{2}$ and the area between 0 and 2.8, namely, $.5000 - .4974 = .0026$.

The second probability is the probability of at most 34 H's in 60 tosses of a coin with $p = .8$. We use the normal approximation with $n = 60$, $p = .8$, $a = 0$, $b = 34$.

Step 1. Solve Eqs. (4.2) for A and B:

$$A = \frac{0 - 60(.8) - .5}{[60(.8)(.2)]^{1/2}} = \frac{-48.5}{3.1} = -15.6 \quad \text{(approximately)},$$

$$B = \frac{34 - 60(.8) + .5}{[60(.8)(.2)]^{1/2}} = \frac{-13.5}{3.1} = -4.4 \quad \text{(approximately)}.$$

Step 2. The area under the standard normal curve between $A = -15.6$ and $B = -4.4$ is, to four decimal places, .0000.

Under the plan of the medical investigator, the claim of the sponsor will be mistakenly accepted when it is false with probability .0026, and mistakenly rejected when valid with probability .0000.   ◁

Example 4.3 *ESP testing.* We recall that a subject in the psychological experiment described in Chapter 2 is considered to have powers of ESP if the probability of his correctly guessing a card is greater than .2. Suppose that a subject is shown 1000 cards. At least how many does he have to guess correctly in order to convince a skeptic that he really has ESP? The skeptic claims that high correct guessing scores are due to "chance": such high scores can appear even though $p = .2$. Let N be a positive integer such that the probability that the Number of H's in 1000 tosses of a coin is at least N is approximately .001 when $p = .2$. The skeptic will be convinced if the number of correct guesses is at least N. What is the value of N with the indicated property?

Step 1. In Table II one sees that the area between $x = 0$ and $x = 3$ under the curve is approximately .4990; thus, the area to the right of $x = 3$ is, by rule 2, equal to the difference $.5000 - .4990 = .001$. We apply the normal approximation with $n = 1000$, $p = .2$, $a = N$, $b = 1000$; A is taken to be equal to 3 and N is unknown.

Step 2. Find N from the first equation in (4.2):

$$N = 1000(.2) + (3)[1000(.2)(.8)]^{1/2} - .5 = 237.42.$$

We conclude: the probability of at least 238 correct guesses out of a total of 1000 is approximately .001; thus, if a subject correctly guesses this many cards, the skeptic will be convinced. Now a subject can correctly guess exactly 200 cards by consistently calling just *one* of the five kinds of cards; thus it is surprising that only 38 more correct guesses have so small a probability.   ◁

Example 4.4 Let us reconsider Example 3.17, in which the poll taker wants to estimate the proportion p of Blue party members in the electorate by the proportion in the sample. Suppose again that the desired margin of error is $d = .01$. It was shown that if a sample of 25,000 voters is selected at random, then the proportion p can be correctly estimated within the margin .01 with probability at least .9; furthermore, a sample of 250,000 will do it with probability at least .99. Now we refine these results by the normal approximation.

If a coin with probability p is tossed 25,000 times, the probability that the frequency of H's is between $p - .01$ and $p + .01$ (which is the same as the probability that the Number of H's is between $25,000p - 250$ and $25,000p + 250$) is, by the normal approximation, equal to the area under the standard normal curve from

$$A = \frac{25,000p - 250 - 25,000p - .5}{(25,000pq)^{1/2}} = -\frac{250 - .5}{50\sqrt{10pq}} = -\frac{1.58}{\sqrt{pq}}$$

to

$$B = \frac{25,000p + 250 - 25,000p + .5}{(25,000pq)^{1/2}} = +\frac{1.58}{\sqrt{pq}}.$$

By virtue of Proposition 3.3, pq does not exceed $\frac{1}{4}$; thus, A does not exceed $-2(1.58) = -3.16$ and B is at least $+3.16$; therefore, the area from A to B is, from Table II, greater than .999. This is larger than .9, obtained by the method of Chapter 3.

Suppose that the poll taker wants the estimate of p to be within the prescribed margin $d = .01$ with probability at least equal to .99; how large a sample size n is sufficient? The probability that the Number of H's in n tosses is between $np - n(.01)$ and $np + n(.01)$ is the area under the standard normal curve between [handwritten: $n = \frac{3^2}{4d^2}$ Take $\frac{1}{2}$ probability, find nearest value in table 2. that = 3.]

$$A = \frac{np - n(.01) - np - .5}{\sqrt{npq}} = \frac{-\sqrt{n}(.01)}{\sqrt{pq}} - \frac{.5}{\sqrt{npq}}$$

and

$$B = \frac{np + n(.01) - np + .5}{\sqrt{npq}} = \frac{\sqrt{n}(.01)}{\sqrt{pq}} + \frac{.5}{\sqrt{npq}}.$$

By virtue of Proposition 3.3, pq is less than or equal to $\frac{1}{4}$; therefore, A is not more than $-2\sqrt{n}(.01) - 2(.5)/\sqrt{n} = -.02\sqrt{n} - .1/\sqrt{n}$, and B not less than $.02\sqrt{n} + .1/\sqrt{n}$; thus, the area under the curve between them is, by rule 7, at least twice the area from 0 to $.02\sqrt{n} + .1/\sqrt{n}$; certainly this is greater than twice the area from 0 to $.02\sqrt{n}$. We note that the area from

0 to 2.58 is .4951; hence, twice the area is .9902. Equating $.02\sqrt{n}$ and 2.58, we find $n = 129^2 = 16,641$. Our conclusion is: the probability that the proportion of members in a sample of $n = 16,641$ differs from the proportion p in the electorate by less than .01 is at least equal to .99; furthermore, 16,641 is the approximate *minimum* sample size for the given margin $d = .01$ and for the given probability .99.   ◁

EXERCISES

1. A coin with probability $p = .3$ is tossed 200 times. Find the probability that the Number of H's is between

a) 66 and 72,   b) 54 and 70,
c) 40 and 50,   d) 55 and 180,
e) 68 and 100,   f) 10 and 70.

2. A coin with probability $p = .6$ is tossed 500 times. Find the probability that the Number of H's is between inclusive

a) 310 and 320,   b) 290 and 315,
c) 280 and 290,   d) 275 and 400,
e) 325 and 390,   f) 200 and 370.

3. A diaper service company sends a blue postcard to the mother of a newborn boy and a pink one to the mother of a newborn girl. Suppose that 1000 babies are born in a community every day. What is the probability that the company will send out more than 540 blue postcards on a given day? Fewer than 475? Between 485 and 510?

4. The natural recovery rate for a certain disease is 50 percent. A new drug is claimed to raise the rate to 75 percent. One hundred patients receive the drug. What is the probability that at least 70 recover if

a) the drug is worthless; or
b) it is really 75 percent effective?

What are the corresponding probabilities for the recovery of 700 out of 1000 patients?

5. Consider a medical hypothesis which states that women have a higher recovery rate from a certain disease than men; more specifically, suppose that the hypothetical rates are 80 percent and 70 percent, respectively. Under the hypothesis, what is the probability that *at least* 74 percent of a sample of 10,000 male patients will recover? What is the probability that *at most* 74 percent of a sample of 10,000 female patients will recover? How can this information be used to *test* the hypothesis?

6. Under the assumption that ESP does not exist, find the probability that a subject correctly guesses at least 125 cards out of a total of 500?

7. A skeptic is willing to accept the existence of ESP if a subject is able to correctly guess a sufficiently large number N out of 10,000 cards. The number N is selected so that the probability of correctly guessing at least N cards is .001 under the assumption of "no ESP" ($p = \frac{1}{5}$); find the value of N.

8. Repeat Exercise 6 for 370 correct guesses out of 1500.

9. Repeat Exercise 7 for a probability of .005.

10. How large a sample of voters is necessary to estimate the proportion of Blue party members with a margin of error of $d = .03$ and with probability at least .98?

11. Repeat Exercise 10 with $d = .02$ and probability at least .96.

12. Repeat Exercise 10 with $d = .005$ and probability at least .999.

chapter 5

ANALYSIS OF THE COIN-TOSSING GAME: EVENTS, CONSISTENCY, AND INDEPENDENCE

In the first four chapters we have developed an elementary theory of the coin-tossing game and have noted its applications. The theory has been entirely concerned with the probability distribution of the Number of H's in n tosses. In this chapter we shall be concerned not only with the Number of H's but also with other results of the game; these will be applied to the "random walk" problem in Chapter 6.

In Section 5.1, events and their probabilities are defined for the coin-tossing model; so are the notions of union, intersection, complement, sure event, and null event. In Section 5.2 it is shown that the assignment of probabilities has an expected consistency property; for example, the probability of H on the first of *two* tosses is the same as the probability of H on the first of 10 tosses. In the final section, the very important idea of *independence* of events is discussed. In Chapters 8 and 9 these results are extended from the coin-tossing model to more general ones.

5.1 EVENTS

Consider the game of tossing a coin n times: the system of outcomes is the system of all multiplets of n letters formed from H and T (Definition 2.1) and the system of probabilities is that in Definition 2.2. There are 2^n such multiplets; indeed, the number of ways of forming a multiplet is the product of the numbers of ways of filling each of the n entries in the multiplet: the first entry can be filled in two ways (H or T), as can each of the $n - 1$ remaining entries, so that the product is $2 \cdot 2 \cdots 2 = 2^n$.

— **Proposition 5.1** *The sum of the probabilities of the outcomes is 1.* —

PROOF. (This has already been proved in the particular cases of 2 and 3 tosses in Chapter 2.)

1. Each of the outcomes may be put into exactly one of $n + 1$ distinct classes according to the number of letters H it contains.

Reason. The first class consists of all outcomes with no letters H, the second consists of all those with exactly one letter H, . . . , and the $(n + 1)$st consists of those with exactly n letters H.

2. The sum of the probabilities of outcomes in each of the above classes is given by the corresponding term of the binomial distribution: the sum of the probabilities of outcomes with exactly k letters H is

$$\binom{n}{k} p^k q^{n-k}.$$

Reason. Definition 2.3 and Proposition 2.3.

3. The sum of the probabilities of the outcomes in all of the above classes is 1.

Reason. By Statements 1 and 2, this is the sum of the terms of the binomial distribution which, by formula (2.6), is equal to 1.   ◄

Definition 5.1 An *event* is a set of outcomes. The probability of an event is defined as the sum of the probabilities of the outcomes in that set.

Example 5.1 The set of outcomes {[HH], [HT]} in formula (2.2) is an event for $n = 2$ tosses; it may be described as the event "H appears on the first of two tosses." Its probability is the sum of the probabilities of [HH] and [HT], respectively: $p^2 + pq = p(p + q) = p.$ ◁

Example 5.2 The set of outcomes {[HTH], [HTT]} is an example of an event for $n = 3$ tosses in (2.3). Its probability is the sum of the probabilities of [HTH] and [HTT], respectively: $p^2 q + q^2 p = pq(p + q) = pq.$

This event may be described as the event "H and T appear on the first and second of three tosses, respectively."   ◁

Events can be combined; for example, if A and B are events, a new event can be formed from them, namely, the event that *both A and B* occur.

Definition 5.2 The *intersection* of two or more events is the event consisting of those outcomes common to both.

Example 5.3 Consider the three events, denoted by the letters A, B, and C, respectively,

$$A = \{[HHH], \quad [HHT], \quad [HTH], \quad [HTT]\},$$
$$B = \{[HTH], \quad [TTT], \quad [TTH], \quad [HTT]\}, \qquad (5.1)$$
$$C = \{[HHT], \quad [TTT], \quad [THT], \quad [HTT]\},$$

from the system of outcomes in formula (2.3). These are the events "H appears on the first toss," "T appears on the second toss," and "T appears on the third toss," respectively. The intersection of A and B is the event consisting of [HTH] and [HTT]: it is the event "H and T appear on the first and second tosses, respectively." The intersections of B and C and of A and C are $\{[TTT], [HTT]\}$ and $\{[HHT], [HTT]\}$, respectively; these are the events "T appears on the last two tosses" and "H appears on the first and T on the third toss," respectively. The intersection of the three events A, B, and C is the event consisting of the single outcome [HTT].   ◁

Events can be combined in another way: if A and B are events, the event that *either A or B or both occur* can be formed from them.

Definition 5.3 The *union* of two or more events is the event consisting of those outcomes belonging to *at least one* of the former events.

Example 5.4 The union of the events A and B in (5.1) consists of the six outcomes:

[HHH], [HHT], [HTH], [HTT], [TTT], [TTH].

The union of A and C consists of six outcomes: those in A as well as [TTT] and [THT]. The union of A, B, and C consists of those in A, and the outcomes [TTT], [TTH], and [THT].

In the display below, the outcomes are classified according to their membership in the events A, B, and C and the intersections and unions;

membership is signified by the letter x under the corresponding event:

outcomes	events			intersections				unions			
	A	B	C	AB	AC	BC	ABC	AB	AC	BC	ABC
HHH	x							x	x		x
HHT	x	x		x				x	x	x	x
HTH	x	x		x				x	x	x	x
THH											
TTH		x						x		x	x
THT			x						x	x	x
HTT	x	x	x	x	x	x	x	x	x	x	x
TTT		x	x			x		x	x	x	x	◁

Definition 5.4 Two events are said to be *disjoint* if they have no common outcomes—in other words, if their intersection contains no outcomes. More generally, several events are called disjoint if no pair of them have outcomes in common.

Example 5.5 The events D and F defined as

$$D = \{[\text{HTH}], [\text{HHT}], [\text{HHH}]\}, \qquad F = \{[\text{TTH}], [\text{THT}], [\text{THH}]\}$$

are disjoint.   ◁

Proposition 5.2 *The probability of the union of two or more disjoint events is equal to the sum of their respective probabilities.*

PROOF. Let A, B, C, ... be disjoint events. Any outcome in their *union* belongs to *one* and *only one* of these, by Definitions 5.3 and 5.4, respectively. It follows that the sum of probabilities defining the probability of their union (Definition 5.1) may be decomposed into (a) the sum of probabilities of outcomes in A, that is, the probability of A, plus (b) the sum of probabilities of outcomes in B, that is, the probability of B, etc.; thus the probability of the union of A, B, C, ... is the sum of the respective probabilities. ◀

There are two simple events of particular importance: the *sure event* consisting of all the 2^n outcomes, and the *null event* containing no outcomes. It follows from Proposition 5.1 and Definition 5.1 that the sure event has probability 1 and the null event has probability 0. The latter event is sometimes called the "impossible" event. Definitions 5.2 and 5.4 imply that two events are disjoint if and only if their intersection is the null event.

Definition 5.5 The complement of an event A is the event consisting exactly of those outcomes not belonging to A.

Example 5.6 The complement of the event A in (5.1) is the event {[THH], [THT], [TTH], [TTT]}. The complement of B is {[HHH], [THT], [THH], [HHT]}.   ◁

Proposition 5.3 *The probability of the complement of an event A is equal to 1 minus the probability of A.*

PROOF

1. An event and its complement are disjoint.

Reason. Definitions 5.4 and 5.5.

2. The probability of A plus the probability of its complement is 1.

Reason. Propositions 5.1 and 5.2.   ◀

Definition 5.6 An event A is said to imply an event B if every outcome in A is also in B; for example, in Example 5.3, the intersection of A and B implies A (and also implies B).

Proposition 5.4 *If one event implies another, then the probability of the former is less than or equal to that of the latter.*

PROOF. Every term in the sum of probabilities giving the probability of the former is also in the sum for the probability of the latter.   ◀

The following proposition will be used in several proofs.

Proposition 5.5 *Let $E_1, \ldots, E_k$ be disjoint events such that their union is the sure event; in other words, every outcome belongs to one and only one of these events. Let A be an event and let $A_1, \ldots, A_k$ be its inter-sections with $E_1, \ldots, E_k$, respectively; that is, each A_i is the inter-section of A and E_i. Then the probability of A is equal to the sum of the probabilities of $A_1, \ldots, A_k$.* ⌐

PROOF

1. Every outcome in A belongs to exactly one of the events $E_1, \ldots, E_k$.

Reason. By assumption.

2. Every outcome in A belongs to exactly one of the events $A_1, \ldots, A_k$.

Reason. This follows from Statement 1 and the definition of the latter sets: an outcome in A which belongs to E_i necessarily belongs to A_i; furthermore, $A_1, \ldots, A_k$ are disjoint.

3. Each of the events $A_1, \ldots, A_k$ implies A.

Reason. By definition.

4. A is equal to the union of $A_1, \ldots, A_k$.

Reason. By Statement 2, A implies the union of $A_1, \ldots, A_k$; by Statement 3, the union of the latter implies A; therefore, the statement follows from the meaning of "implies" (Definition 5.6).

5. The assertion of the proposition follows.

Reason. Statement 4 and Proposition 5.2.   ◀

Example 5.7 Let A be as in (5.1). Let $E_1, \ldots, E_4$ be the following events:

$$E_1 = \{[HHH], [THH]\}, \qquad E_2 = \{[HHT], [THT]\},$$
$$E_3 = \{[HTH], [TTH]\}, \qquad E_4 = \{[HTT], [TTT]\}.$$

The latter events satisfy the condition in the hypothesis of Proposition 5.5: they are disjoint and their union is the sure event. The intersections $A_1, \ldots, A_4$ of A with $E_1, \ldots, E_4$ are

$$A_1 = \{[HHH]\}, \qquad A_2 = \{[HHT]\},$$
$$A_3 = \{[HTH]\}, \qquad A_4 = \{[HTT]\};$$

clearly A is the union of $A_1, \ldots, A_4$, and the probability of A is the sum of the probabilities of the latter.   ◁

We close this section with two observations.

Definition 2.3 can be considered a particular case of Definition 5.1. Take the event "exactly k letters H turn up in n tosses" as the set of outcomes containing exactly k letters H. The probability of this event is the same as the "probability that the Number of H's is equal to k" as given by Definition 2.3.

A reader with some knowledge of the theory of sets must have noticed that the definitions and propositions in this section are identical with those in set theory; one has only to substitute the terms "points" and "sets" for the terms "outcomes" and "events," respectively. Instead of recording (another) exposition of the abstract theory of sets, and *then* making the appropriate substitution of terms, I thought it better to employ the terminology of probability in the *development* of the theory.

5.1 EXERCISES

1. Enumerate the system of outcomes in the particular case $n = 4$. Carry out each step of the proof of Proposition 5.1 in this particular case.

2. Consider the events (for $n = 4$):

$$A = \{[HTHH], [HTHT], [HTTH], [HTTT]\},$$
$$B = \{[HHTH], [HTTH], [THTH], [TTTH]\},$$
$$C = \{[HTTH], [HTTT], [TTTH], [TTTT]\},$$
$$D = \{[THHH], [THHT], [THTH], [THTT]\}.$$

Describe each of the following events, and find their probabilities:

a) the unions of all combinations of these events taken two, three, and four at one time;
b) the intersections of all such combinations;
c) the complements of A, B, C, and D.

3. Repeat Exercise 2 for the events

$$E = \{[HHHH], [HHTH], [HTHH], [HTTH]\},$$
$$F = \{[HHHH], [HHHT], [HHTH], [HHTT]\},$$
$$G = \{[HHHT], [HHTT], [HTHT], [HTTT]\},$$
$$H = \{[TTHH], [TTHT], [TTTH], [TTTT]\}.$$

4. Prove the following statement: The intersection of A and B implies A.

5. Prove: A implies the union of A and B.

6. Prove: If A implies B, then the complement of B implies the complement of A.

7. (Refer to Exercises 2 and 3.) Let E_1, E_2, E_3, and E_4 be the events defined as $E_1 = A$, $E_2 = D$, $E_3 = F$, and $E_4 = H$. Show that this collection of events satisfies the conditions in the hypothesis of Proposition 5.5.

8. A coin is tossed four times. Let A be the event "two H's appear before two T's" and B the event "T appears at least once in the first two tosses." Enumerate the outcomes in A and B, respectively, their intersection, unions, and respective complements. Find the corresponding probabilities when $p = \frac{1}{2}$.

9. A coin is tossed three times. Let E_1, E_2, and E_3 be the events that H occurs for the first time on the first, second, and third tosses, respectively, and E_4 the event that H does not turn up at all. Let A be the event "two successive H's appear somewhere in the course of the three tosses." Assuming $p = \frac{2}{3}$, find the probability of A by applying Proposition 5.5.

10. In a family with five children, what is the probability that a third boy is born before a second girl?

5.2 CONSISTENCY

The event A in (5.1) consists just of those outcomes for three tosses which have H on the first toss; hence, A may be described as the event "H appears on the first of three tosses." The event B consists of those outcomes for three tosses which have T on the second toss: it is the event "T appears on the second of the three tosses." Similarly, C is the event "T appears on the third of three tosses." The intersection of A and B may be described as the event "H and T occur on the first and second of the three tosses, respectively."

Our system of probabilities has a fundamental property of *consistency*, now to be described. The probability of A in (5.1) is, in accordance with Definition 5.1, equal to

$$p^3 + p^2q + p^2q + pq^2 = p^2(p + q) + pq(p + q)$$
$$= p^2 + pq = p(p + q) = p;$$

in other words, the event "H appears on the first of three tosses" has probability p. The event in Example 5.1, "H on the first of two tosses," was shown to have probability p. Finally, the outcome H on a single toss also has the same probability p; hence, the probability of H on the first toss is the same for one, two, or three tosses. This is a particular case of a general principle which we now formally introduce.

Definition 5.7 In the system of outcomes of n tosses of a coin, an event of the form "H (or T) appears on the first toss, . . . , H (or T) appears on the kth toss" is called an event *determined* by the first k tosses. More generally, an event "H (or T), H (or T), . . . appear on specified tosses" is an event determined by those tosses.

Example 5.8 The events A, B, and C in (5.1) are determined by the first, second, and third tosses, respectively; for example, an outcome belongs to B if and only if T appears on the *second* toss. The intersection of A and

B is determined by the first and second tosses because it is of the form "H and T appear on the first and second tosses, respectively."   ◁

⌐ **Consistency Theorem.** *The event "H on the first of n tosses" has the same probability p for any number n of tosses. More generally, an event determined by a set of tosses has the same probability in any game of n tosses which includes the given set of tosses.* ⌡

PROOF. We shall prove only the first statement; the proof of the second is conceptually similar to that of the first but requires more notation. Let us show that the event "H on the first of n tosses" has the same probability as "H on the first of $n - 1$ tosses." This is sufficient for the proof: since we have already seen that the probability is p for 1, 2, or 3 tosses, the previous statement implies that the probability is p for $n = 4, 5, \ldots$ tosses.

1. The event "H on the first of $n - 1$ tosses" consists of all multiplets of $n - 1$ letters H and T which start with the letter H.

Reason. Use $n - 1$ in place of n in Definition 2.1.

2. The event "H on the first of n tosses" consists of all multiplets of n letters starting with H.

Reason. Definition 2.1.

3. With every outcome in the event in Statement 1 we can associate exactly two outcomes in the event in Statement 2 in this way: the first $n - 1$ letters in each of the latter coincide with the letters in the former, while the nth-place letters are H and T, respectively; for example, with the $(n - 1)$-letter outcome $[H \cdots T]$ we associate the two n-letter outcomes $[H \cdots TH]$ and $[H \cdots TT]$. All the n-letter outcomes can be "paired" in such a manner.

Reason. Each multiplet of $n - 1$ letters H and T can be extended in exactly two ways to a multiplet of n letters by adjoining H or T as the nth letter.

4. The probabilities of the n-letter outcomes in Statement 3 obtained by adjoining H or T to an $(n - 1)$-letter outcome are equal to the probability of the latter outcome multiplied by p or q, respectively.

Reason. Definition 2.2.

5. The sum of the probabilities of two n-letter outcomes obtained by adjoining H and T, respectively, to an $(n - 1)$-letter outcome is equal to the probability of that $(n - 1)$-letter outcome.

Reason. By Statement 4, the common factor in the two former probabilities is the latter probability; when this common factor is removed, the sum of the remaining factors is $p + q = 1$.

6. The *sum* of the probabilities of the outcomes in the event "H on the first of $n - 1$ tosses" is equal to the sum of the probabilities of the outcomes in the event "H on the first of n tosses."

Reason. By Statement 5, each term in the former sum is the sum of exactly two terms of the latter sum; conversely, by Statement 3, the terms of the latter sum can be *paired* so that each pair has a sum equal to a particular term of the former sum.

7. The events "H on the first of $n - 1$ tosses" and "H on the first of n tosses" have the same probability.

Reason. Statement 6 and Definition 5.1. ◄

5.2 EXERCISES

1. (Refer to Exercises 2 and 3, Section 5.1.) By which tosses are the events $A, B, \ldots , H$ determined?

2. Which tosses determine the following events (from Exercise 1): the intersections of B and C, A and E?

3. Prove: If two events are determined by the same set of tosses, then they are disjoint or identical.

4. Carry out the details of the proof of the Consistency Theorem in the particular case $n = 4$.

5.3 INDEPENDENCE

The concept of independence, itself describable in elementary terms, is found in almost every part of the theory of probability and its applications. Most of the fundamental theoretical results depend on this concept in a basic way. The notion of independence has been implicit in all that has been covered in Chapters 2 and 3; now we formulate it for use in the remaining chapters.

Definition 5.8 Two events are called *independent* if the probability of their intersection is equal to the product of their respective probabilities. More generally, several events are called independent if the probability of the intersection of any number of them is equal to the product of the corresponding probabilities; for example, three events A_1, A_2, and A_3 are independent if the probability of their intersection is equal to the product of the probabilities of A_1, A_2, and A_3, and if the probability of the intersections of A_1 and A_2, of A_1 and A_3, and of A_2 and A_3 are equal to the products of the probabilities of A_1 and A_2, A_1 and A_3, and A_2 and A_3, respectively.

Example 5.9 The three events A, B, and C in (5.1) are independent. The intersections are described in Example 5.3. The probabilities of A, B, and C are p, q, and q, respectively. The probabilities of the intersections of A, B, and C, of A and B, of A and C, and of B and C are pq^2, pq, pq, and q^2, respectively.   ◁

The events A, B, and C in (5.1) are determined by different tosses in the sense of Definition 5.6; they are also independent. This is a particular case of a general principle:

Independence Theorem. Two or more events determined by different tosses, or by nonoverlapping sets of different tosses, are independent; for example, if an event is determined by the first k out of n tosses, and another is determined by the last $n - k$ tosses, $k < n$, then the two events are independent.

PROOF. We shall give the proof of just the particular case of the *two* events mentioned. The proof of the general case embodies the same ideas but is notationally more complicated; hence, we omit it. Let A and B be events determined by the first k and by the last $n - k$ tosses, respectively.

1. If A is an event determined by the first k tosses and B an event determined by the last $n - k$ tosses, then the intersection of A and B is determined by all n tosses; that is, it consists of a single outcome.

Reason. This follows directly from Definition 5.7: the intersection of A and B is necessarily of the form "H (or T) appears on the first toss, . . . , H (or T) appears on the nth toss."

2. The probability of A is a product of k factors p and q; the probability of B is a product of $n - k$ factors p and q; and the probability of their intersection is the product of all n factors p and q.

Reason. Consistency Theorem.

3. The events A and B are independent.

Reason. Statement 2 and Definition 5.8. ◄

The Independence Theorem can be extended from events determined by particular tosses to *unions* of such events; for example, in a game of n tosses, let A be the event "exactly one H appears on the first two tosses," and B the event "exactly two H's appear on the second two tosses"; then A is a union of the two events determined by the first two tosses, "H and T appear on the first and second tosses, respectively," and "T and H appear on the first and second tosses, respectively," and B is the event "H appears on the third and fourth tosses." The events A and B are independent. The intersection of A and B is the union of the two events "H, T, H, H appear on the first, . . . , fourth tosses, respectively," and "T, H, H, H appear on the first, . . . , fourth tosses, respectively"; hence, by the Consistency Theorem, its probability is $2qp^3$. By virtue of that theorem, the probabilities of A and B are $2qp$ and p^2, respectively; therefore, A and B are independent.

We shall use the following fact about events:

Proposition 5.6 *Let $A', A'', \ldots, B$ be events, A the union of $A', A'', \ldots,$ and $C', C'', \ldots$ the intersections of A' and B, A'' and B, $\ldots$, respectively; then*

$$\text{intersection of } A \text{ and } B = \text{union of } C', C'', \ldots.$$

PROOF. We shall show that an outcome belongs to one of these if and only if it belongs to the other.

1. An outcome belongs to the intersection of A and B when (i) it belongs to at least one of the sets $A', A'', \ldots$; and (ii) it also belongs to B.

Reason. Hypothesis and definition.

2. An outcome belongs to the union of $C', C'', \ldots$ when it belongs to at least one of the intersections A' and B, A'' and B, ...

Reason. Definition of $C', C'', \ldots$.

3. The assertion of the proposition follows.

Reason. An outcome satisfies the condition in Statement 1 if and only if it satisfies the condition in Statement 2. ◄

Example 5.10 Let A' be the event "H and T appear on the first and second tosses, respectively," and A'' the event "T and H appear on the first and second tosses, respectively"; then their union A is the event "exactly one H appears in the first two tosses." Let B be the event "H appears on the third and fourth tosses"; and let C' and C'' be the intersections of A' and B, and A'' and B, respectively. We shall show that the intersection of A and B is the union of C' and C''. Now C' is the event "H, T, H, H appear on the first, . . . , fourth tosses, respectively," and C'' the event "T, H, H, H appear on the first, . . . , fourth tosses, respectively"; their union is just the intersection of A and B. ◁

The following proposition implies the extension of the Independence Theorem to *unions* of events determined by particular tosses.

Proposition 5.7 *Let A', A'', . . . , B be events such that A', A'', . . . are disjoint and such that A' and B, A'' and B, . . . are pairs of independent events; let A be the union of A', A'', . . . ; then A and B are independent.*

PROOF

1. If, in Proposition 5.6, the events A', A'', . . . are disjoint, then so are C', C'', ...

Reason. (See Definition 5.4.) If there is no outcome common to the pair A' and A'', then there is surely none common to their intersections with B.

2. The probabilities of C', C'', . . . are the products of the probabilities of A' and B, A'' and B, . . . , respectively.

Reason. By hypothesis, the pairs A' and B, A'' and B, . . . are pairs of independent events.

3. The probability of the union of C', C'', . . . is the sum of the products of the probabilities of A' and B, A'' and B, . . . , respectively.

Reason. Statement 2 and Proposition 5.2.

4. The sum of the products of the probabilities of A' and B, A'' and B, . . . is equal to the probability of A multiplied by the probability of B.

Reason. The probability of B is the common factor of these products. When it is removed, the sum of the remaining factors is the sum of the probabilities of A', A'', ... This sum is the probability of (their union) A by Proposition 5.2.

5. The probability of the intersection of A and B is the product of their respective probabilities.

Reason. By Proposition 5.6, the probability of the intersection of A and B is the probability of the union of C', C'', . . . ; the latter probability, by Statements 3 and 4, is equal to the product of the probabilities of A and B.

6. The assertion of the proposition follows.

Reason. Statement 5 and Definition 5.7.   ◀

Example 5.11 (Refer to Example 5.10.) The event A is the union of the two events A' and A'', which are disjoint. A' and B are independent, as are A'' and B, by virtue of the Independence Theorem. The probabilities of C' and C'' are both p^3q. The intersection of A and B has probability $2p^3q$, which is the product of the probabilities $2pq$ and p^2 of A and B, respectively. ◁

The following proposition shows that the independence relation among events implies the same relation among their complements; even more:

Proposition 5.8 Let A_1, . . . , A_m be independent events, and C_1, . . . , C_m their respective complements. The independence property of the former aggregate is preserved when any of its members is replaced by the corresponding complement: each collection $(C_1, A_2, . . . , A_m)$, $(A_1, C_2, A_3, . . . , A_m)$, $(C_1, C_2, A_3, . . . , A_m)$, . . . , $(C_1, C_2, . . . , C_m)$ consists of independent events.

PROOF. Let a_1, . . . , a_m be the probabilities of A_1, . . . , A_m, respectively; consequently, $1 - a_1$, . . . , $1 - a_m$ are the probabilities of C_1, . . . , C_m, respectively. For the proof, it is sufficient to show that the independence of A_1, . . . , A_m implies that of C_1, A_2, . . . , A_m because the independence relation is *symmetric* with respect to the labeling of the events. We have to show that the probability of the intersection of *any* subcollection $(C_1, A_2, . . . , A_k)$, $k \le n$, of $(C_1, A_2, . . . , A_m)$ is the product of the probabilities $(1 - a_1)a_2 \cdots a_k$.

1. Let A_1, . . . , A_k be k events selected from among A_1, . . . , A_m, $k \le m$. The probability of their intersection is the product of their probabilities a_1, . . . , a_k.

Reason. Definition of independence (Definition 5.8).

2. The probability of the intersection of A_2, . . . , A_k is the sum of the probabilities of the intersections: (i) of A_1, A_2, . . . , A_k, and (ii) of C_1, A_2, . . . , A_k.

Reason. The events A_1 and C_1 are disjoint and their union is the sure event, because one is the complement of the other. The intersection of $A_2, \ldots, A_k$ has the intersection (i) with A_1 and the intersection (ii) with C_1; hence, the statement follows from Proposition 5.5.

3. The probability of the intersection of $C_1, A_2, \ldots, A_k$ is the product $a_2 \cdots a_k$ minus the product $a_1 a_2 \cdots a_k$.

Reason. Note first that Statement 1 itself implies that the probability of the intersection of $A_2, \ldots, A_k$ is the product $a_2 \cdots a_k$; thus, our statement is a consequence of Statements 1 and 2.

4. The probability of the intersection of $C_1, A_2, \ldots, A_k$ is the product of the respective probabilities $(1 - a_1)a_2 \cdots a_k$.

Reason. Statement 3 and the algebraic identity

$$a_2 \cdots a_k - a_1 a_2 \cdots a_k = (1 - a_1)a_2 \cdots a_k. \qquad \blacktriangleleft$$

Example 5.12 A coin is tossed n times. Let A_1 be the event "H on the first toss" and C_1 its complement, "T on the first toss." Let A_2 be the event "H on both the second and third tosses" and C_2 its complement, "at least one T on the second and third tosses." By the Independence Theorem, A_1 and A_2 are independent because the former is an event determined by the first toss and the latter determined by the second and third tosses. It follows from Proposition 5.8 that the following pairs of events are independent: A_1 and C_2, A_2 and C_1, C_1 and C_2. $\triangleleft$

5.3 EXERCISES

1. Let $A_1, A_2, \ldots, A_n$ be the events "H appears on the first toss," $\ldots$, "H appears on the nth toss," respectively; and let $D_1, D_2, \ldots, D_n$ be the events "T appears on the first toss," $\ldots$, "T appears on the nth toss," respectively. Describe each of the following events:

a) the intersection of A_1, D_2, and A_4,
b) the union of A_1, A_2, and D_2,
c) the union of A_n and D_n,
d) the intersection of A_3 and D_3.

2. Describe the event "exactly 2 H's appear in the first three tosses" as a union of intersections of sets $A_1, A_2, \ldots, D_1, D_2, \ldots$.

3. Consider the three events:

$$E_1 = \text{intersection of } A_1, D_2, \text{ and } D_6,$$
$$E_2 = \text{intersection of } A_3 \text{ and } A_5,$$
$$E_3 = \text{intersection of } D_4 \text{ and } A_7.$$

Find the probabilities of E_1, E_2, and E_3 and of all their intersections; then verify that they are independent.

4. Let F be the union of E_2 and E_3. Prove: E_1 and F are independent. Let G be the union of E_1 and E_3; are F and G independent?

5. Let A be the event "exactly one H appears in the first three tosses"; B the event "at most one H occurs in the second three tosses"; and C the event "at least one H occurs in the third set of three tosses." Prove that A, B, and C are independent.

6. Consider three events such that every *two* of them are independent; are all *three* events independent? The following example furnishes the answer. A balanced coin is tossed twice. Let A be the event "H on the first toss," B the event "H on the second toss," and C the event "exactly one H appears in the two tosses." Show that every *two* of these are independent, but not all *three*.

7. What is the maximal number of independent events for two tosses of a coin; in other words, what is the largest number of events which, as a group, are independent?

8. A coin is tossed 30 times. Let A_1, A_2, and A_3 be the events that at least one H appears in the first 10, second 10, and third 10 tosses, respectively. Show that these three are independent events.

9. A coin is tossed five times. A "tie" is said to occur on the kth toss if the Number of H's on the first k tosses is equal to the Number of T's. (Ties can occur only on even-numbered tosses.) Let A be the event "at least one tie occurs before the fifth toss," and B the event "H appears on the fifth toss." Show that A and B are independent. (Hint: Express A as a union of disjoint events which are independent of B, and apply Proposition 5.7.)

10. Here is a converse to Proposition 5.8: If $A_1, \ldots, A_m$ are events such that

i) the probability of their intersection is the product of their respective probabilities, and

ii) the condition just stated remains true when any number of the events are replaced by their respective complements,

then $A_1, \ldots, A_m$ are independent. Prove this for the particular case of three events A_1, A_2, A_3.

chapter 6

RUNS AND THE RANDOM WALK

When a coin is tossed several times, it may happen that H appears on a number of consecutive tosses. Such an occurrence is called a "run" of H's of the specified length. The main result of Section 6.1 is contained in Proposition 6.3: it shows that for a fixed number r, if a game is allowed to go for a large number of tosses, it is highly probable that a run of r consecutive H's will occur at some point during the game. This result is applied to the "Ruin Principle" in Section 6.2: if two gamblers start with finite quantities of capital and bet a fixed amount on each toss of a coin, and if the game is allowed to continue for a very long time, then sooner or later one of the players will be "ruined," that is, will lose everything he started with. The main result of Section 6.3 is the derivation of the probabilities of ruin for each of the players; this is applied to sequential hypothesis testing in Chapter 13.

6.1 RUNS

The object of this section is the determination of the magnitude of the probability that a run of r consecutive H's occurs somewhere in the course of n tosses. It is shown that, if r is fixed and n is allowed to increase indefinitely, then this probability approaches 1. Since a simple form for this probability is not obtainable, we are content to show that it is at least as great as some associated but simpler quantity which approaches 1 under the stated conditions.

Proposition 6.1 *If the coin is tossed n times, the probability that at least one H appears is* $1 - q^n$; *thus, the limit of this sequence of probabilities for* $n = 1, 2, \ldots$ *is equal to 1 (Definition 3.5).*

PROOF

1. The probability of the event "no H's occur in the course of n tosses" is q^n.

Reason. This event consists of the single outcome "T on all n tosses," which has probability q^n.

2. The event "at least one H appears" is the complement of the event "no H's appear."

Reason. Definition 5.5.

3. The probability of the event "at least one H appears" is $1 - q^n$.

Reason. The probability of its complement is q^n by Statements 1 and 2; this and Proposition 5.3 imply the statement.

4. The limit of the sequence of numbers $1 - q, 1 - q^2, \ldots, 1 - q^n, \ldots$ is 1.

Reason. The differences between 1 and the terms of the sequence form the sequence $q, q^2, \ldots, q^n, \ldots$. This has the limit 0 because q is less than 1.

Statements 3 and 4 contain the assertions of the proposition.   ◄

We now introduce the coin-tossing game with "block tossing." Suppose the integer n is a product of two positive integers m and r; that is, $n = mr$. We divide the n tosses into m blocks of r tosses each: the first block consists of tosses 1 through r, the second consists of tosses $r + 1$ through $2r$, etc. Let $A_1, \ldots, A_m$ be the events

$$A_1 = \text{"at least one T appears in the first block,"}$$
$$A_2 = \text{"at least one T appears in the second block," etc.,}$$

(6.1)

and let $C_1, \ldots, C_m$ be their respective complements.

Proposition 6.2 *The probability of the intersection of* $A_1, \ldots, A_m$ *is* $(1 - p^r)^m$.

PROOF

1. The complements $C_1, \ldots, C_m$ are independent.

Reason. The event C_1 is "H appears on each of the first r tosses"; hence, it is determined by the first r tosses (Definition 5.8). In the same way, one sees that $C_2, \ldots, C_m$ are determined by the second, $\ldots$, mth blocks of r tosses, respectively. The statement now follows from the Independence Theorem (Section 5.3).

2. The events $A_1, \ldots, A_m$ are independent.

Reason. The complements of independent events are independent (Proposition 5.8).

3. Each of the events $A_1, \ldots, A_m$ has probability $1 - p^r$.

Reason. Apply the first part of Proposition 6.1 to each A, reversing the respective roles of H and T, p and q, and replacing n by r.

The proposition follows from Statements 2 and 3, and the definition of independence (Definition 5.8). ◀

Suppose a coin is tossed several times. A *run* of (consecutive) H's (or T's) is an uninterrupted string of H's (or T's); for example, if the coin is tossed five times, the outcome [THHHT] contains a *run* of three H's.

Definition 6.1 A run of r H's (or T's) is said to be completed at the kth toss if H (or T) turns up on the kth toss and on each of the preceding $r - 1$ tosses, while T (or H) turns up on the $(k - r)$th toss. A run is said to occur during the game if the run is completed on one of the n tosses.

Example 6.1 A run of three T's is completed on the eighth toss if H turns up on the fifth and T on the sixth, seventh, and eighth tosses. A run of five H's is completed on the ninth toss if T occurs on the fourth and H on the fifth through ninth tosses. ◁

The following proposition is a generalization of Proposition 6.1; one of its conclusions is that, if a coin is tossed a very large number of times, the probability is very close to 1 that a run of r H's (or T's) will occur.

Proposition 6.3 *Let r be a positive integer. A coin is tossed n times; let m be the largest multiple of r less than or equal to n, that is, the integral part of the quotient n/r. Then the probability of at least one run of r H's* *is greater than or equal to*

$$1 - (1 - p^r)^m.$$

For fixed r, the limit of the sequence of probabilities for n = 1, 2, ... is 1. The probability of at least one run of r T's is at least equal to

$$1 - (1 - q^r)^m;$$

and for fixed r, the limit of the corresponding sequence of probabilities is 1.

PROOF. The proof is based on the game with "block tossing" and the previous propositions. Since mr is less than or equal to n, we may divide the first mr tosses into m blocks of r tosses each.

1. The event "no run of r H's appears in the course of n tosses" implies the event A defined as the intersection of $A_1, \ldots, A_m$ in formula (6.1).

Reason. If no run of r H's appears, then each block of r tosses must produce at least one T.

2. The probability of the former event in Statement 1 is less than or equal to $(1 - p^r)^m$, which is the probability of A.

Reason. Proposition 5.4, Statement 1, and Proposition 6.2.

3. The probability of the complement of A is not greater than the probability of the event "at least one run of r H's appears in the course of n tosses."

Reason. Statement 2 and Proposition 5.3.

4. The probability of the event "at least one run of r H's appears in the course of n tosses" is greater than or equal to $1 - (1 - p^r)^m$.

Reason. Statements 2 and 3.

5. The limit of the sequence of probabilities of "at least one run of r H's appears in the course of n tosses" for $n = 1, 2, \ldots$ is 1.

Reason. The difference between 1 and the probability for n tosses is less than or equal to $(1 - p^r)^m$ by Statement 4. As n increases, so does m (since the latter is defined as the largest multiple of r less than n). Since p is less than 1, so is $1 - p^r$; therefore, the sequence $(1 - p^r)^m$, $m = 1, 2, \ldots$ has the limit 0 in accordance with Definition 3.3; therefore, the sequence of probabilities in Statement 5 has the limit 1 in accordance with Definition 3.5.

Statements 4 and 5 complete the proof of the part of Proposition 6.3 concerning runs of H's. The part about T's follows by interchanging the respective roles of H and T, and p and q.   ◄

Example 6.2 A coin is tossed 250 times; let us use Proposition 6.3 to find a lower bound on the probability of getting a run of four H's. Here we have $n = 250$ and $r = 4$; therefore, m is the integral part of the quotient $\frac{250}{4}$, or 62. The quantity $1 - (1 - p^r)^m$ is $1 - (1 - p^4)^{62}$; for example, if $p = \frac{1}{2}$, it is approximately equal to $1 - .018 = .982$; therefore, *if a fair coin is tossed* 250 *times, the probability is at least* .982 *that there will be a run of four* H's.

Let us now estimate the probability of a run of seven H's in 250 tosses; here, $n = 250$ and $r = 7$, so that $m = 35$. The lower bound on the probability is $1 - (1 - p^7)^{35}$. In the particular case where $p = \frac{1}{2}$, this lower bound is approximately equal to .5.   ◁

6.1 EXERCISES

1. Consider the game of $n = 5$ tosses. Enumerate the outcomes forming each of the events:
a) there is a run of three H's,
b) there is no run of two H's,
c) there is no run of four T's.

2. A balanced coin is tossed 100 times. Using Proposition 6.3, estimate the probability of a run of three H's; of four H's.

3. Repeat Exercise 2 with $p = .8$ and runs of six H's and two T's.

4. Where, in the proof of Proposition 6.2, has the Consistency Theorem (Section 5.2) been tacitly used?

5. How has the concept of independence been essential to the proof of Proposition 6.3?

6.2 THE RANDOM WALK

Suppose that two gamblers play the following simple coin-tossing game. A coin with probability p of H is repeatedly tossed. Each time H turns up, the first player, whom we shall call Abel, collects one dollar from the second player, whom we shall call Cain. Each time T turns up, Abel pays Cain one dollar. After the first toss, Abel either has a gain of one dollar or a loss of one dollar: his *fortune* is either $+1$ or -1. After the second toss, his fortune is either $+2$, 0, or -2; and, after the third toss, it is either $+3$,

$+1$, -1, or -3. The *record of his fortune* in three tosses is a description of the fortune at each toss: each of the eight outcomes of the three tosses corresponds to exactly one of eight records of fortune:

outcome	record	probability
[HHH]	[1, 2, 3]	p^3
[HHT]	[1, 2, 1]	p^2q
[HTH]	[1, 0, 1]	p^2q
[THH]	[−1, 0, 1]	p^2q
[TTT]	[−1, −2, −3]	q^3
[TTH]	[−1, −2, −1]	q^2p
[THT]	[−1, 0, −1]	q^2p
[HTT]	[1, 0, −1]	q^2p

As an example, consider the outcome [HTH]: Abel wins one dollar on the first toss, loses it on the second, and wins it back on the third; thus, his record of fortune is described as [1, 0, 1]. The *actual* record is a variable depending on the Outcome of the three tosses: we refer to it as the "Record of Fortune (with capital letters) in three tosses."

Definition 6.2 A record of fortune corresponding to an outcome of n tosses of a coin is a multiplet of n integers such that

a) the first integer is either $+1$ or -1, depending on whether H or T is the first letter in the outcome; and

b) each integer after the first is one unit larger or smaller than its predecessor, depending on whether H or T is the corresponding letter in the outcome.

The Record of Fortune (capital letters) in n tosses is the variable depending on the Outcome of n tosses.

The Record of Fortune is also called a "random walk." Imagine that the players keep score by marking units 0, ± 1, ± 2, . . . on a straight line, placing a particle at 0, and moving it one unit in the positive or negative direction for each H or T, respectively, that turns up. The particle is said to undergo a "random walk" on the integers.

Suppose that the players start the game each with an initial capital, and agree to play until either one of the players loses all his capital or n tosses have been completed; in the former case, we say that the losing player has been "ruined." One of the fundamental problems in the theory of the random walk is the "ruin problem": it is to determine the probability that a particular player is ruined.

Example 6.3 Consider a game of three tosses in which Abel starts with two dollars and Cain with one; let us compute the probabilities that Abel loses, Cain loses, and that neither loses. Abel loses if he has a net loss of two dollars before Cain has a net loss of one. There are two records with the property: $[-1, -2, -3]$ and $[-1, -2, -1]$, corresponding to the outcomes [TTT] and [TTH]; hence, the event "Abel loses" has probability $q^3 + q^2p = q^2$. Cain loses if he has a net loss of one dollar before a net gain of two. There are five records with this property: $[1, 2, 3]$, $[1, 2, 1]$, $[1, 0, 1]$, $[-1, 0, 1]$, $[1, 0, -1]$, corresponding to the outcomes [HHH], [HHT], [HTH], [THH], and [HTT], respectively; hence, the event "Cain loses" has probability $p^3 + 3p^2q + pq^2$. Neither player loses for the record $[-1, 0, -1]$ corresponding to the outcome [THT]; hence, the probability that neither player is ruined is q^2p (see Fig. 6.1).   ◁

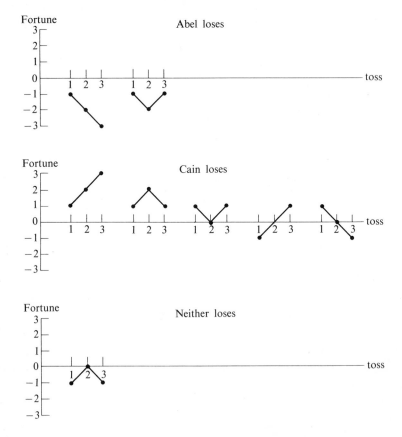

Fig. 6.1. Records of Fortune for game of three tosses in which Abel starts with two dollars and Cain with one.

√ ⌐*The Ruin Principle.* Let R_n be the probability that one of the players
is ruined before or at the completion of n tosses. Then the limit of the
sequence $R_1, R_2, \ldots, R_n, \ldots$ is 1; in other words, if a game is allowed
to continue for many tosses, there is a large probability (nearly 1) that
one of the players will have been ruined by the time the game had been
scheduled to end.⌐

PROOF. The ruin principle is a consequence of the theory of runs in
the previous section. Let us denote by $r + 1$ the sum of the initial capitals
of the two players. If n is larger than $r + 1$, and if a run of r H's occurred,
then one of the players must have been ruined. This may be seen as follows.
If neither player had been ruined at the time of the toss starting the run
of r H's, then Abel certainly must have had one dollar; therefore, *Cain
could not have had more than r dollars*; hence, Cain must have *lost at least
r dollars* during the run; thus, he must have been ruined. It follows from this
discussion that the event "a run of r H's appears" implies the event "one
of the two players is ruined"; hence, by Proposition 5.4, the probability
of the latter is at least as large as that of the former. Proposition 6.3 states
that the probability of the former event is close to 1 if n is large; hence, the
probability of the latter event has the same property. ◄

Now we shall show that, if the number n of tosses is very large, then
the probability that Cain (or Abel) is ruined is practically independent of the
number n; in other words, the probability is about the same for all sufficiently large n. Let E_n, F_n, and G_n stand for the following events:

$$E_n = \text{"Cain is ruined in a game of } n \text{ tosses,"}$$
$$F_n = \text{"Abel is ruined in a game of } n \text{ tosses,"} \qquad (6.2)$$
$$G_n = \text{"neither player is ruined in a game of } n \text{ tosses."}$$

Let e_n, f_n, and g_n be their respective probabilities. We first prove the following
preliminary proposition.

⌐*Proposition 6.4* The sequence of numbers $(e_1, e_2, \ldots, e_n, \ldots)$,
$(f_1, f_2, \ldots, f_n, \ldots)$, and $(g_1, g_2, \ldots, g_n, \ldots)$ have the following
properties:

a) $e_1 \leq e_2 \leq \cdots \leq e_n \leq \cdots$,

b) $f_1 \leq f_2 \leq \cdots \leq f_n \leq \cdots$,

c) $e_1 + f_1 + g_1 = 1, \ldots, e_n + f_n + g_n = 1, \ldots$. ⌐

PROOF

1. $e_{n-1} \le e_n$ for every n. This proves (a).

Reason. e_{n-1} is the probability that Cain is ruined in the course of $n - 1$ tosses. The latter event implies E_n, defined in (6.2); hence, by Proposition 5.4, we have $e_{n-1} \le e_n$.

2. $f_{n-1} \le f_n$.

Reason. This is proved by the same argument as for Statement 1: substitute Abel for Cain.

3. $e_n + f_n + g_n = 1$.

Reason. The events E_n, F_n, and G_n are disjoint, and their union is the sure event; therefore, Proposition 5.2 implies the statement.   ◀

Here is the formal statement of the result on the probabilities of ruin:

Proposition 6.5 *The three sequences in Proposition 6.4 have limits e_0, f_0, and 0, respectively; in other words, the probabilities that Cain and Abel are ruined are very nearly equal to fixed numbers e_0 and f_0, respectively, for all large n; and the probability that neither is ruined is very nearly 0.*

PROOF

1. The sequence $g_1, g_2, \ldots, g_n, \ldots$ has the limit 0.

Reason. The event G_n in (6.2) is the complement of the event "one of the players is ruined before or at the completion of n tosses"; hence, by the Ruin Principle and Proposition 5.3, the sequence of probabilities corresponding to the events $G_1, G_2, \ldots, G_n, \ldots$ has the limit 0.

2. There exists a real number e_0 representing the limit of the sequence $e_1, e_2, \ldots, e_n, \ldots$.

Reason. Each successive term in the sequence is at least as large as its predecessor (Proposition 6.4), and each is less than or equal to 1 because it is a probability. It follows that the sequence satisfies the hypothesis of the statement at the end of Section 3.2 with $b = 1$ (the consequence of the completeness property of the real number system); hence, the sequence has a limit (cf. Section 3.2, Exercise 8).

3. There exists a real number f_0 representing the limit of the sequence $f_1, f_2, \ldots, f_n, \ldots$.

Reason. Analogous to Statement 2.   ◀

6.2 EXERCISES

1. Cain and Abel play a game of $n = 4$ tosses. Enumerate the outcomes in the events:

i) Cain is ruined before or at the completion of the tosses,
ii) event (i) with Abel in place of Cain,
iii) neither is ruined before or at the completion of the tosses,

when Cain starts with one dollar and Abel with four.

2. Repeat Exercise 1 for the case where Cain starts with three dollars and Abel with two.

3. In a game of five tosses of a balanced coin, what are the probabilities of the events in Exercise 1 for each division of a total capital of four between the players?

4. The proof of the Ruin Principle gives more information than is claimed in its statement: If $r + 1$ represents the sum of the initial capitals of the two players, then the probability that one of the players is ruined before or at the completion of n tosses is at least $1 - (1 - p^r)^m$, where m is the largest multiple of r less than or equal to n. Demonstrate this.

5. Use the result of Exercise 4 to prove that the probability e_n in Proposition 6.4 differs from the limit e_0 by no more than $(1 - p^r)^m$.

6.3 SOLUTION OF THE RUIN PROBLEM

In this section we assume that the number of tosses n is very large. Suppose that Cain and Abel have a combined capital c of which Cain initially has r dollars, where $r < c$; what is the probability that Cain loses? Let us call this sought probability $w_r^{(n)}$. According to Proposition 6.5, the terms of the sequence $w_r^{(1)}, w_r^{(2)}, \ldots, w_r^{(n)}, \ldots$ (which were called $e_1, e_2, \ldots$ when the subscripts and superscripts were omitted) have a limit which will be denoted here as w_r. The latter will be referred to simply as "the probability that Cain loses"—without reference to the number of tosses. The terms of the sequence of probabilities $w_r^{(1)}, w_r^{(2)}, \ldots, w_r^{(n)}, \ldots$ with sufficiently

Cain bets on T $P(H) = p$
Abel bets on H

large superscripts n are all very close to w_r; hence, in the calculations that follow, we shall use w_r as an approximation to $w_r^{(n)}$ when n is very large, deriving linear equations in the quantities w_r. (In other words, if n is very large, the probability that Cain is ruined in the course of a game of n tosses is practically independent of the number n, and so the index n may be suppressed.)

The "probability that Cain loses," w_r, cannot be conveniently found by summation of probabilities of outcomes, as the exercises of Section 6.2 indicate. The classical method of finding it is the method of simultaneous difference equations. Before the formal derivation in the general case, we present a particular example to illustrate the method.

Suppose the total capital is $c = 3$. Let w_1 and w_2 be the probabilities that Cain is ruined when he starts with $r = 1$ and $r = 2$ dollars, respectively. For purposes of convenience, we also define w_0 as 1 and w_3 as 0. These definitions are consistent with the model: Cain has probability 1 of being ruined if he "starts" with zero dollars, and probability 0 of being ruined if he "starts" with three dollars. We shall show that the quantities w_0, w_1, w_2, and w_3 satisfy the following system of simultaneous equations:

$$w_1 = pw_0 + qw_2, \qquad w_2 = pw_1 + qw_3. \qquad (6.3)$$

Since $w_0 = 1$ and $w_3 = 0$, the system has two unknowns, w_1 and w_2; the solution is

$$w_1 = p/(1 - qp), \qquad w_2 = p^2/(1 - qp).$$

Let us outline the derivation of the first equation in (6.3). The event "Cain is ruined when he starts with one dollar to Abel's two" may be decomposed into two disjoint events, depending on the outcome of the first toss. Let A be the event "H on the first toss," and B the event "T on the first toss"; A and B are disjoint events whose union is the sure event. The event that Cain is ruined when he starts with one dollar is the union of the two disjoint events:

a) the intersection of the events "Cain is ruined when he starts with one dollar" and A, and

b) the intersection of the events "Cain is ruined when he starts with one dollar" and B.

In the first case, when Cain starts with one dollar, and when H occurs on the first toss, his capital is cut to 0; the probability for this is p. In the second case, when he starts with one dollar and T occurs on the first toss, then his capital is increased to two dollars. At the start of the second toss he is in the position of starting with an initial capital of two dollars: his probability of being ruined when he starts on the *second* toss with two dollars is the same as if he had started on the *first* toss with two dollars, because the probability is practically independent of the number of tosses in the game (Proposition 6.5); furthermore, the event B and the event that Cain is ruined when he starts on the second toss with two dollars are independent because of the Independence Theorem and Proposition 5.7; therefore, the probability of the *second* intersection mentioned above is qw_2. The first equation in (6.3) is a statement of the fact that w_1, the probability of ruin starting with one dollar, is the sum of the probabilities p (or pw_0) and qw_2 of the disjoint events displayed above. The argument for the second equation in (6.3) is analogous. Here is a complete statement and proof of the general case.

Proposition 6.6 *Let* $w_0, \ldots, w_r, \ldots, w_c$ *be the probabilities that Cain is ruined when he starts with initial capitals* $0, \ldots, r, \ldots, c$, *respectively; we put* $w_0 = 1$ *and* $w_c = 0$. *The above probabilities satisfy the system of equations:*

$$w_r = qw_{r+1} + pw_{r-1}, \qquad r = 1, \ldots, c - 1. \qquad (6.4)$$

PROOF. Let A, B, K, L, and M be the following events:

A = "H on the first toss,"

B = "T on the first toss,"

K = "Abel's Record of Fortune reached r before ever falling to $r - c$" (this is the same as the event that Cain is ruined when starting with initial capital r),

L = "the successive net gains of Abel in the tosses *after* the first reached $r - 1$ before ever falling to $r - 1 - c$,"

M = "the successive net gains of Abel in the tosses after the first reached $r + 1$ before ever falling to $r + 1 - c$."

For the moment we shall assume $1 < r < c - 1$.

1. The intersection of A and K is equal to the intersection of A and L; and the intersection of B and K is equal to the intersection of B and M.

Reason. This is a consequence of the definitions of these events.

2. The probabilities of K, L, and M are w_r, w_{r-1}, and w_{r+1}, respectively.

Reason. The probability of K is, by *definition*, equal to w_r. The event L is equivalent to "Cain loses when starting with initial capital $r - 1$" in a game in which the *first* toss is not counted; hence, by Proposition 6.5, its probability is w_{r-1} because the probability is independent of the number of tosses. The reasoning for the probability of M is the same.

3. The events A and L are independent; so are the events B and M.

Reason. A is determined by the first toss, and L is a union of disjoint events determined by the other tosses; hence, by Proposition 5.7, they are independent. The reasoning for B and M is similar.

4. The probabilities of the intersections of A and L and of B and M are pw_{r-1} and qw_{r+1}, respectively.

Reason. Statements 2 and 3.

5. The probability of K is the sum of the probabilities of the intersections of A and K and of B and K, respectively.

Reason. A and B are disjoint events whose union is the sure event; therefore, the statement is a consequence of Proposition 5.5.

6. Equation (6.4) is valid for $1 < r < c - 1$.

Reason. Statements 1 and 4. ◀

We omit the details of the proof in the cases $r = 1$ and $r = c - 1$. They are almost the same as the proof given above if the definitions of L and M are modified as follows: define L as the *sure event* if $r = 1$ and M as the *null event* if $r = c - 1$. The only change necessary in the previous proof is the *reason* for Statement 3: we use the fact that two events are independent if one of them is the sure event or the null event; this is a simple consequence of previous definitions.

Example 6.4 Suppose that the sum of the initial capitals of the players is $c = 5$. The quantities $w_1, \ldots, w_4$ satisfy the system of simultaneous equations (where $w_0 = 1$, $w_5 = 0$):

$$
\begin{aligned}
w_1 &= pw_0 + qw_2 = p + qw_2, \\
w_2 &= pw_1 + qw_3, \\
w_3 &= pw_2 + qw_4, \\
w_4 &= pw_3 + qw_5 = pw_3.
\end{aligned}
\tag{6.5}
$$

The system can be solved by the standard method of successive elimination of the unknowns. Replace w_1 in the second equation by $p + qw_2$ from the first; and replace w_4 in the third equation by pw_3 from the last; thus, one obtains a system of two equations in two unknowns:

$$w_3(1 - pq) = pw_2, \qquad w_2(1 - pq) = qw_3 + p^2,$$

whose solution is

$$w_3 = p^3/[(1 - pq)^2 - pq], \qquad w_2 = p^2(1 - pq)/[(1 - pq)^2 - pq].$$

The unknowns w_1 and w_4 are now obtained directly from the first and last of the equations (6.5). ◁

Example 6.5 A coin-tossing game of particular interest is the one with a balanced coin, that is, with $p = q = \frac{1}{2}$; the random walk in this case is called a "symmetric" random walk, because at each toss there is equal probability that the Record of Fortunes rises or falls by a unit. The system of equations (6.4) takes the form

$$w_1 = (\tfrac{1}{2})(1 + w_2), \qquad w_2 = (\tfrac{1}{2})(w_1 + w_3), \qquad \dots,$$
$$w_{c-2} = (\tfrac{1}{2})(w_{c-3} + w_{c-1}), \qquad w_{c-1} = (\tfrac{1}{2})w_{c-2}. \qquad (6.6)$$

The unique solution of this system is

$$p = q = \tfrac{1}{2} \qquad w_1 = (c - 1)/c, \qquad w_2 = (c - 2)/c, \qquad \dots,$$
$$w_{c-2} = 2/c, \qquad w_{c-1} = 1/c; \qquad (6.7)$$

thus, *the probability that Cain is ruined is equal to the proportion of the total capital initially in the hands of Abel.* One consequence of this is that in gambling with a fair coin a poor man is likely to be ruined by a rich opponent. ◁

There is a general formula for the unique solution of the system of equations (6.4) when p and q are not equal:

$$w_r = \frac{(p/q)^c - (p/q)^r}{(p/q)^c - 1} . \qquad p \neq q \qquad (6.8)$$

This is the only solution of the system (6.4) because it has *at most one* solution; the proof is outlined in Exercise 10. Formula (6.8) does provide a solution: the equations $w_0 = 1$ and $w_c = 0$ are certainly fulfilled while for $0 < r < c$ Eqs. (6.4) are also satisfied:

$$qw_{r+1} + pw_{r-1} = q \cdot \frac{(p/q)^c - (p/q)^{r+1}}{(p/q)^c - 1} + p \cdot \frac{(p/q)^c - (p/q)^{r-1}}{(p/q)^c - 1}$$
$$= \frac{(q + p)(p/q)^c - (p + q)(p/q)^r}{(p/q)^c - 1} = w_r.$$

* concerning p : see bottom of p. 74

Example 6.6 For $c = 4$ and $p = .2$ we have $p/q = \frac{1}{4}$, and so the solution (w_1, w_2, w_3) is

$$w_1 = \tfrac{63}{255}, \qquad w_2 = \tfrac{15}{255}, \qquad w_3 = \tfrac{3}{255}. \qquad \triangleleft$$

6.3 EXERCISES

1. Cain and Abel play a game with many tosses, with a total capital of $c = 5$. Write down the equations for the probabilities of ruin. Find the solution in the particular case $p = .4$.

2. Repeat Exercise 1 for $c = 5$, $p = .6$. What relation is there to the probabilities obtained in Exercise 1?

3. Repeat Exercise 1 for $c = 6$, $p = .2$.

4. The probabilities w_r in the system (6.4) are limits of the probabilities of ruin in a game of n tosses. Compare the probabilities of ruin obtained for $n = 5$ in Exercise 3, Section 6.2, with the corresponding probabilities obtained as the solution of the system (6.4).

5. Complete the details of the proof of Proposition 6.6 in the case $r = 1$.

6. Repeat Exercise 5 in the case $r = c - 1$.

7. In the reason for Statement 3 in the proof of Proposition 6.6, it is claimed that the event L is a union of disjoint events determined by the tosses following the first. Show the truth of this claim in the case of a game with at most five tosses, $r = 2$, $c = 4$.

8. Repeat Exercise 7 for the event M defined in the proof.

9. The proof of the Ruin Principle in Section 6.2 and, in particular, the result of Exercise 4 following it, provide a quantitative uniform bound on the deviation between the *exact* probability of ruin in a game of n tosses and the *limit* probability obtained from the system (6.4). Prove: Let m be the largest multiple of $c - 1$ not exceeding n; then the deviation just mentioned does not exceed $(1 - p^{c-1})^m$. The proof is based on Proposition 6.4 and the exercise noted above.

10. Prove that the system (6.4) has at most one solution, as follows:

a) Let $(u_0, u_1, \ldots, u_c)$ and $(v_0, v_1, \ldots, v_c)$ be two solutions; show that the condition $u_0 = v_0 = 1$ implies $u_1 - v_1 = q(u_2 - v_2)$.

b) Show that the result in (a) implies that $u_2 - v_2 = (u_3 - v_3)q/(1 - pq)$.

c) Extend the result in (b): prove that either $u_r = v_r$ or $|u_r - v_r| < |u_{r+1} - v_{r+1}|$.

d) Starting with the condition $u_c = v_c = 0$ and working from the other end of the system (6.4), show, by analogy to (b) and (c), that either $u_r = v_r$ or $|u_{r+1} - v_{r+1}| < |u_r + v_r|$.

e) The conclusions of statements (c) and (d) imply the identity of the two solutions.

chapter 7

RANDOM VARIABLES FOR COIN TOSSING

A "function," in mathematical terminology, is a rule of association between two sets, the domain and the range, such that to each element in the domain there corresponds exactly one in the range. Functions arising in algebra and calculus usually have sets of real numbers as domain and range; such functions can be represented geometrically by their graphs. A major part of probability theory is about a very different kind of function: one whose domain is a system of outcomes of a game and whose range is a set of real numbers. Such a function is called a "random variable"; this term is suggested by the observation that the "values are determined by chance, or randomness." Such a function cannot be represented by a graph on a two-dimensional plane; instead, it is characterized by its "probability distribution." In this chapter we study random variables for the coin-tossing game; several applications, including the theory of "group testing," are presented. Random variables for more general kinds of games are introduced in Chapter 8.

7.1 THE CONCEPT OF A RANDOM VARIABLE

Consider the game of tossing a coin n times. A random variable for this game is a *numerical representation* of the system of outcomes: it is a rule (or "function") which associates a real number with each outcome of the game.

Example 7.1 Consider the rule which associates a number with each outcome in a game of three tosses as follows:

outcome	number	outcome	number
[HHH]	3	[TTT]	0
[HHT]	2	[TTH]	1
[HTH]	2	[THT]	1
[THH]	2	[HTT]	1

This is just the rule that associates the number of H's in the outcome with the outcome. It is convenient to identify this random variable (rule) as the "Number of H's in three tosses": we are using the same term for both the *rule* and the actual number of H's.   ◁

Example 7.2 Consider the rule of association, or random variable, described as follows:

outcome	number	outcome	number
[HHH]	3	[TTT]	0
[HHT]	2	[TTH]	1
[HTH]	1	[THT]	1
[THH]	2	[HTT]	1

This is the rule which associates with each outcome the length of the longest run of H's in the outcome; for example, there is a run of two H's in the outcome [HHT]. We call this random variable the "longest run of H's in three tosses."   ◁

Example 7.3 *Doubling the odds.* Cain plays Abel a game of three tosses with the following rules:

a) One dollar is bet on the first toss: if T appears, Cain collects one dollar from Abel and quits; if H appears, he pays Abel one dollar and bets two dollars on the second toss.

b) If T appears on the second toss, Cain collects two dollars from Abel and quits (with a net gain of one dollar); if H appears, he pays Abel two more dollars and bets four dollars on the third toss.

c) Cain wins or loses four dollars on the third toss accordingly as the outcome of the third toss is T or H, respectively.

Consider Cain's net gain in the three tosses as a random variable which we call "Cain's gain in three tosses":

outcome	net gain	outcome	net gain
[HHH]	−7	[TTT]	1
[HHT]	1	[TTH]	1
[HTH]	1	[THT]	1
[THH]	1	[HTT]	1

◁

Definition 7.1 The *set of values* of a random variable is the set of numbers in the corresponding numerical representation of the outcomes.

The set of values of the random variable "Number of H's in three tosses" (Example 7.1) consists of the integers 0, 1, 2, and 3; the set of values of the random variable "longest run of H's in three tosses" (Example 7.2) consists of the same set of integers; and the set of values of the random variable "Cain's gain in three tosses" consists of the integers −7 and +1.

It is customary to use capital letters X, Y, Z, etc., to stand for random variables, and lower-case letters x, y, z, etc., for the numbers in the sets of values.

Definition 7.2 Let X be a random variable and x a number in its set of values. The set of outcomes numerically represented by x is called the "event $\{X = x\}$."

Let X be the Number of H's in three tosses. The event $\{X = 0\}$ consists of the single outcome [TTT]; the event $\{X = 1\}$ consists of the outcomes [TTH], [THT], and [HTT]; the event $\{X = 2\}$ consists of the outcomes [HHT], [HTH], and [THH]; and the event $\{X = 3\}$ consists of the single outcome [HHH].

Let Y be the longest run of H's in three tosses; the events $\{Y = y\}$ are:

event	consists of outcomes
$\{Y = 0\}$	[TTT]
$\{Y = 1\}$	[HTH], [TTH], [THT], [HTT]
$\{Y = 2\}$	[HHT], [THH]
$\{Y = 3\}$	[HHH]

Let Z be Cain's gain in three tosses; then the event $\{Z = -7\}$ consists of the single event [HHH] and the event $\{Z = 1\}$ consists of the other seven outcomes.

Definition 7.3 Let us denote by $\Pr(X = x)$ the probability of the event $\{X = x\}$. Suppose that $x_1, x_2, \ldots$ are the values of X; then the set of values $x_1, x_2, \ldots$, together with the system of probabilities $\Pr(X = x_1)$, $\Pr(X = x_2), \ldots$, is called the *probability distribution* of X.

The probability distribution of the random variable X in Example 7.1 is given by

$$\Pr(X = 0) = q^3, \qquad \Pr(X = 1) = 3q^2p,$$
$$\Pr(X = 2) = 3qp^2, \qquad \Pr(X = 3) = p^3;$$

this is the binomial distribution with $n = 3$.

The probability distribution of the random variable Y in Example 7.2 is given by

$$\Pr(Y = 0) = q^3, \qquad \Pr(Y = 1) = 3q^2p + p^2q,$$
$$\Pr(Y = 2) = 2p^2q, \qquad \Pr(Y = 3) = p^3.$$

The probability distribution of the random variable Z in Example 7.3 is given by

$$P\{Z = -7\} = p^3, \qquad P\{Z = 1\} = 1 - p^3.$$

Proposition 7.1 *The sum of the terms in the probability distribution of a random variable is 1.*

PROOF. Let X be a random variable with the set of values $x_1, x_2, \ldots$. The events $\{X = x_1\}, \{X = x_2\}, \ldots$ are disjoint events whose union is the sure event; in other words, every outcome in the system belongs to exactly one of these events. The assertion of the proposition now follows from Proposition 5.2. ◄

Since a random variable has *several* numbers in its set of values, it is useful to define an "average" of these values. The average we shall use is a weighted average: each value x is weighted by the corresponding term $\Pr(X = x)$ in the probability distribution. We have already introduced this kind of average in Chapter 3: the Expected Number of H's. Now we generalize.

Definition 7.4 The *expected value* of a random variable X is a weighted average of its values: if $x_1, \ldots, x_k$ are the values of X, the expected value of X, denoted $E(X)$, is defined as

$$E(X) = x_1 \Pr(X = x_1) + \cdots + x_k \Pr(X = x_k).$$

Let X, Y, and Z be the random variables given in Examples 7.1, 7.2, and 7.3, respectively; then

$$E(X) = (0)q^3 + (1)3q^2 p + (2)3qp^2 + (3)p^3,$$

which is $3p$, as calculated in Section 3.1.

$$E(Y) = (0)q^3 + (1)3pq + (2)p^2 q + (3)p^3$$
$$= 3p - 3p^2 + 2p^2 - 2p^3 + 3p^3 = 3p - p^2 + p^3,$$
$$E(Z) = -7p^3 + (1 - p^3) = 1 - 8p^3.$$

Now we extend the notion of *variance* first mentioned in Chapter 3; it is a measure of the "spread" of the values of a random variable about its expected value.

Definition 7.5 Let X be a random variable with the set of values $x_1, \ldots, x_k$. The *variance* of X, denoted by $\mathrm{Var}(X)$, is defined as the sum

$$\mathrm{Var}(X) = [x_1 - E(X)]^2 \Pr(X = x_1) + \cdots + [x_k - E(X)]^2 \Pr(X = x_k).$$

It may be verbally described as the weighted average of the squares of the differences between the values of X and $E(X)$; each squared difference $[x - E(X)]^2$ is weighted by the corresponding probability $\Pr(X = x)$. If there is a large probability $\Pr(X = x)$ for a value x for which the squared difference $[x - E(X)]^2$ is large, then the product $[x - E(X)]^2 \cdot \Pr(X = x)$ contributes a large term to the sum defining the variance; on the other hand, if either $\Pr(X = x)$ is very small and $[x - E(X)]^2$ is of moderate size, or the latter is very small and the former is of moderate size, then the product contributes little to the sum defining the variance.

As an illustration, let us compute the probability distribution, expected value, and variance of the random variable Y in Example 7.2 in the particular case $p = \frac{1}{2}$:

$$\Pr(Y = 0) = \tfrac{1}{8}, \quad \Pr(Y = 1) = \tfrac{1}{2}, \quad \Pr(Y = 2) = \tfrac{1}{4}, \quad \Pr(Y = 3) = \tfrac{1}{8};$$
$$E(Y) = (0)(\tfrac{1}{8}) + (1)(\tfrac{1}{2}) + (2)(\tfrac{1}{4}) + (3)(\tfrac{1}{8}) = \tfrac{11}{8};$$
$$\mathrm{Var}(Y) = [0 - \tfrac{11}{8}]^2 \cdot \tfrac{1}{8} + [1 - \tfrac{11}{8}]^2 \cdot \tfrac{1}{2} + [2 - \tfrac{11}{8}]^2 \cdot \tfrac{1}{4}$$
$$+ [3 - \tfrac{11}{8}]^2 \cdot \tfrac{1}{8} = \tfrac{47}{64}.$$

The corresponding quantities for the random variable Z in Example 7.3, for $p = \frac{1}{2}$, are

$$\Pr(Z = -7) = \tfrac{1}{8}, \qquad \Pr(Z = 1) = \tfrac{7}{8};$$
$$E(Z) = (-7)\tfrac{1}{8} + 1 \cdot \tfrac{7}{8} = 0;$$
$$\mathrm{Var}(Z) = (-7)^2 \cdot \tfrac{1}{8} + (1)^2 \cdot \tfrac{7}{8} = 7.$$

7.1 EXERCISES

1. Let Y be the random variable for a game of four tosses which associates with each of the 16 outcomes the length of the longest run of H's in the outcome; for example, the longest run of H's in the outcome [HHHT] is 3. Find the set of values of Y, its probability distribution, expected value, and variance in the particular case $p = .5$.

2. Let Z be the random variable representing Cain's net gain in a game of *four* tosses in which the odds are doubled after each H (see Example 7.3). Find the set of values of Z, its probability distribution, expected value, and variance in the particular case $p = .5$.

3. Let X be the random variable for a game of three tosses with the following numerical representation of the outcomes:

outcome	number	outcome	number
[HHH]	2	[TTT]	0
[HHT]	2	[TTH]	0
[HTH]	1	[THT]	1
[THH]	1	[HTT]	1

Find the set of values of X, its probability distribution, expected value, and variance in terms of the quantity p.

4. Let X be a random variable for a game of four tosses with the following numerical representation of the outcomes: each outcome for which the number of T's exceeds the number of H's is represented by the integer 1, and each outcome in which the number of T's does *not* exceed the number of H's is represented by the integer 0. Find the probability distribution, expected value, and variance of X in the case $p = .5$.

5. Let X be a random variable whose set of values consists of the single number c; this is the "constant random variable." Show that $E(X) = c$ and $\mathrm{Var}(X) = 0$.

6. For any number b and random variable X, let bX be the random variable obtained from X by multiplying each of its values by b. Prove: $E(bX) = bE(X)$ and $\mathrm{Var}(bX) = b^2 \mathrm{Var}(X)$.

7. For a number c and a random variable X let $X + c$ be the random variable obtained from X by adding c to each of its values. Prove: $E(X + c) = E(X) + c$ and $\text{Var}(X + c) = \text{Var}(X)$.

8. Show that the previous three exercises imply:

$$E(bX + c) = bE(X) + c, \qquad \text{Var}(bX + c) = b^2 \, \text{Var}(X).$$

7.2 AN ILLUSTRATION: "DOUBLING THE ODDS" IN A GAME OF n TOSSES

Consider the coin-tossing game of Chapter 6: Abel wins one dollar from Cain each time H appears and loses one dollar to him each time T appears. Suppose that the coin is balanced, that is, that $p = q = \frac{1}{2}$. Let X be the random variable representing the net gain of Abel after n tosses. For simplicity let us assume that n is an even integer; thus, the net gain is either $0, +2, -2, \ldots,$ or $+n, -n$. The expected value $E(X)$ is equal to 0; this can be seen as follows. By definition, $E(X)$ is the sum

$$0 \cdot \Pr(X = 0) + 2 \cdot \Pr(X = 2) + (-2)\Pr(X = -2) + \cdots$$
$$+ n\Pr(X = n) + (-n)\Pr(X = -n).$$

The event $\{X = 2\}$ has the same probability as the event $\{X = -2\}$ because the coin is balanced; in other words, the outcomes represented by a net gain of $+2$ for Abel have the same probabilities as those represented by a net gain of $+2$ for Cain; thus $\Pr(X = 2) = \Pr(X = -2)$; hence, the second and third terms in the sum for $E(X)$ cancel each other. For the same reason all succeeding pairs of terms in the sum add up to 0; therefore, the sum itself is 0: $E(X) = 0$. The game is considered "fair" because each player has an "expected net gain" equal to 0.

Consider the following modification of the game: Cain is permitted to double the odds after each H and to withdraw after the first T, as in the particular case $n = 3$ treated in Example 7.3. Let Z be the random variable representing Cain's net gain in this game; let us calculate the probability distribution and expected value of Z. If all n tosses produce H's, Cain's loss is $1 + 2 + 2^2 + \cdots + 2^{n-1}$ because the loss on each toss is twice that on the preceding toss. The sum $1 + 2 + \cdots + 2^{n-1}$ is the sum of a finite geometric series with ratio 2: it is equal to $(1 - 2^n)/(1 - 2)$, or $2^n - 1$. The event $\{Z = -2^n + 1\}$ is the event "H appears on all n tosses," and so has probability 2^{-n}. If at least one T appears in the course of the n tosses, then Cain's net gain is 1; indeed, if he wins for the first time on the kth toss, he wins 2^{k-1} for *that* toss, but has already lost

$$1 + 2 + \cdots + 2^{k-2} = 2^{k-1} - 1$$

on the previous $k - 1$ tosses for a net gain of $2^{k-1} - (2^{k-1} - 1) = +1$.

The event "Cain wins on at least one of the tosses" is the complement of the event "H appears on all n tosses," and so, by Proposition 5.3, has probability $1 - 2^{-n}$. The result of these calculations is that the probability distribution of Z is

$$\Pr(Z = -2^n + 1) = 2^{-n}, \qquad \Pr(Z = +1) = 1 - 2^{-n}.$$

It follows that

$$E(Z) = (-2^{-n} + 1)2^{-n} + (+1)(1 - 2^{-n}) = 0;$$

thus, the "expected net gain" of Cain is 0, so that the game is "fair."

7.3 APPLICATION: TOURNAMENT PROBABILITIES

Suppose Cain and Abel play the following coin-tossing game. The coin is successively tossed and the game ends when either the fourth H or fourth T first appears: Abel wins if the fourth H appears first, and Cain wins if the fourth T appears first. The game can be considered to have seven tosses; if either player wins before the seventh toss, then the results of the remaining tosses are disregarded.

Let N be the random variable representing the number of tosses necessary to determine the winner; that is, it is the number of the toss on which the fourth H or T first appears. The set of values of N is the set of integers 4, 5, 6, 7; we shall calculate the probability distribution of N. First,

$$\Pr(N = 4) = p^4 + q^4.$$

Reason. The event $\{N = 4\}$ consists of the two outcomes "H appears on each of the first four tosses" and "T appears on each of the first four tosses," which have probabilities p^4 and q^4, respectively. Our next calculation is

$$\Pr(N = 5) = \binom{4}{3} p^4 q + \binom{4}{3} q^4 p.$$

Reason. Let E_1, E_2, A, B, A_1, and B_2 be the events:

$$E_1 = \text{"H on the fifth toss,"}$$
$$E_2 = \text{"T on the fifth toss,"}$$
$$A = \text{"three H's in the first four tosses,"}$$
$$B = \text{"three T's in the first four tosses,"}$$
$$A_1 = \text{intersection of } A \text{ and } E_1,$$
$$B_2 = \text{intersection of } B \text{ and } E_2.$$

The following statements are valid:

1. A and E_1 are independent.

Reason. A is a union of disjoint events determined by the first four tosses; hence, the statement follows from Proposition 5.7.

2. B and E_2 are independent.

Reason. Same as for Statement 1.

3. The probability of A_1 is $\binom{4}{3}p^4q$.

Reason. The probabilities of A and E_1 are $\binom{4}{3}p^3q$ and p, respectively; hence, by Statement 1, their intersection A_1 has the indicated probability.

4. The probability of B_2 is $\binom{4}{3}pq^4$.

Reason. The probabilities of B and E_2 are $\binom{4}{3}pq^3$ and q, respectively; hence, by Statement 1, their intersection B_2 has the indicated probability.

5. $\Pr(N = 5)$ has the value indicated above.

Reason. The event $\{N = 5\}$ is the union of the two disjoint events A_1 and B_2; hence, by Proposition 5.2, its probability is the sum of the probabilities in Statements 3 and 4.

This completes the computation of $\Pr(N = 5)$. The formula

$$\Pr(N = 6) = \binom{5}{3}p^4q^2 + \binom{5}{3}p^2q^4$$

is derived in an analogous way; the reader is asked to carry out the derivation (see Exercise 1 below). The probability $\Pr(N = 7)$ is $\binom{6}{3}p^3q^3$; indeed, the event $\{N = 7\}$ is equivalent to the event "three H's appear in the first six tosses."

In the particular case of a balanced coin, the probability distribution of N is

$$\Pr(N = 4) = \tfrac{1}{8}, \quad \Pr(N = 5) = \tfrac{1}{4}, \quad \Pr(N = 6) = \tfrac{5}{16}, \quad \Pr(N = 7) = \tfrac{5}{16}.$$

(It is often said that a World Series involving two evenly matched teams "most probably" lasts seven games; however, using the above model, we find that it is just as likely to last six games.)

The "expected number of tosses necessary to determine the winner," or $E(N)$, is $E(N) = 4 \cdot \tfrac{1}{8} + 5 \cdot \tfrac{1}{4} + 6 \cdot \tfrac{5}{16} + 7 \cdot \tfrac{5}{16} = 5\tfrac{13}{16}$; thus, $E(N)$ is less than 6.

A general formula for the expected number of tosses in a tournament will be derived in Chapter 13; there it is applied to the theory of testing hypotheses.

7.3 EXERCISES

1. Complete the derivation of the formula for $\Pr(N = 6)$.

2. Compute the probability distribution of N, and $E(N)$ and $\text{Var}(N)$ in the particular cases $p = .4, .3$.

3. Derive the probability distribution of N in the case of a tournament which is completed after the fifth H or T.

4. Repeat Exercise 2 for the case considered in Exercise 3.

7.4 APPLICATION TO GROUP TESTING

During World War II large numbers of men were called for service and had to be medically tested for the presence of certain diseases. Some of these could be detected from a small sample of a man's blood. (A test on the sample indicates whether or not the agents of the disease are present.) Since so many men were examined, a large number of individual blood samples had to be tested; this would have taken much time and cost much money. The method of *group testing* was introduced.

In the *group testing* method, the blood samples of groups of men are pooled and a single test is made on the pooled sample: if the disease agents are not found in it, then none of the donors have the disease, and only one test is required; on the other hand, if the agents are found in the pooled sample, then at least one of the donors has the disease, and each of the blood samples has to be individually retested. In the latter case, the number of tests required is one greater than the number of donors to the pooled sample.

We shall use the coin-game model to analyze the group testing method. Suppose that the presence or absence of the disease in an individual is determined by a (hypothetical) coin toss: H and T correspond to presence and absence, respectively. By the Law of Large Numbers, the probability p of H represents the proportion of the population having the disease. We are implicitly assuming that the presence of the disease in one man does not influence its presence in others. Suppose that the blood samples are pooled in groups of size k. The probability that the agents of the disease are absent from a pooled sample is q^k; and the probability that the agents are present is, by Proposition 5.3, equal to $1 - q^k$.

Let X be the number of tests necessary to test a group of k blood samples; X has the set of values 1, $k + 1$. The event $\{X = 1\}$ is equivalent to the event that the agents are *absent* from the pooled sample; hence,

$$Pr(X = 1) = q^k.$$

The event $\{X = k + 1\}$ is the event that the agents are *present* in the pooled sample; hence,

$$Pr(X = k + 1) = 1 - q^k.$$

It follows from these two formulas that

$$E(X) = q^k + (k + 1)(1 - q^k) = k + 1 - kq^k.$$

If the blood samples are individually tested (that is, without group testing), then k tests are *always* necessary to determine the presence or absence of the disease in each group of men. The difference between k and the "expected number of tests under group testing" is

$$k - E(X) = k - [k + 1 - kq^k] = kq^k - 1.$$

If p is small, that is, if the disease is rare, and k is not too large, then the above difference $kq^k - 1$ is of significant size; for example, if $p = .01$ and $k = 5$, then

$$kq^k - 1 = 5(.99)^5 - 1 = 3.76 \text{ (to two decimal places)};$$

thus, the "expected saving" on the number of tests is approximately 3.76 for every five men, a saving of about 75 percent.

The table and figure on page 96 are reproduced from the article by R. Dorfman, "The Detection of Defective Members in Large Populations," in *Annals of Mathematical Statistics*, Vol. 14 (1943), pp. 436–440. In Table 7.1, the "Prevalence rate (percent)" is $100p$; the "Optimum group size" is the value of k which maximizes the relative saving

$$\frac{k - E(X)}{k} = q^k - \frac{1}{k};$$

the "Relative testing cost" is the corresponding minimum of $100 \cdot E(X)/k$; and the "Percent saving attainable" is the difference between 100 and the

**Table 7.1   Optimum Group Sizes and Relative Testing Costs
for Selected Prevalence Rates**

Prevalence rate (percent)	Optimum group size	Relative testing cost	Percent saving attainable
1	11	20	80
2	8	27	73
3	6	33	67
4	6	38	62
5	5	43	57
6	5	47	53
7	5	50	50
8	4	53	47
9	4	56	44
10	4	59	41
12	4	65	35
13	3	67	33
15	3	72	28
20	3	82	18
25	3	91	9
30	3	99	1

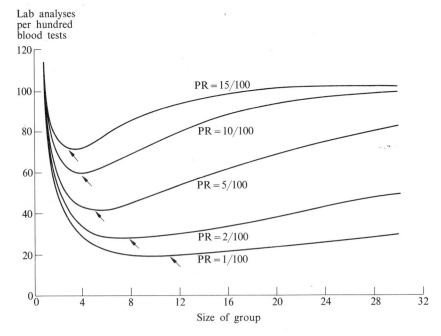

Fig. 7.1. Economies resulting from blood testing by groups (PR denotes prevalence rate).

latter. Note that the saving is greatest when p is small, least when p is large. In Fig. 7.1 the quantity $100 \cdot E(X)/k$, as a function of k, is traced on a graph for various prevalence rates (PR).

7.5 COMPOSITE RANDOM VARIABLES AND JOINT PROBABILITY DISTRIBUTIONS

More than one random variable can be defined for a game of n tosses of a coin; for example, three random variables were presented in Examples 7.1, 7.2, and 7.3, respectively, for the game of three tosses. Consider two random variables representing the outcomes of a game; when taken together, they form a *composite* random variable which associates a *pair* of numbers with each outcome of the game: the two numbers forming the pair are values assigned by the individual random variables to the particular outcome.

> **Definition 7.6** Let X_1 and X_2 be random variables representing the outcomes of a game. The composite random variable formed from X_1 and X_2 is a representation of the outcomes by *pairs* of numbers: with each outcome we associate the pair of numbers assigned by X_1 and X_2, respectively, to that particular outcome. The composite random variable is denoted by $[X_1, X_2]$.

Example 7.4 Let X_1 and X_2 be the random variables in Examples 7.1 and 7.2, respectively. The composite random variable $[X_1, X_2]$ assigns pairs of numbers to the outcomes of the three tosses in accordance with the following rule:

outcome	number pair	outcome	number pair
[HHH]	[3, 3]	[TTT]	[0, 0]
[HHT]	[2, 2]	[TTH]	[1, 1]
[HTH]	[2, 1]	[THT]	[1, 1]
[THH]	[2, 2]	[HTT]	[1, 1] ◁

Example 7.5 Let X_1 and X_3 be the random variables in Examples 7.1 and 7.3, respectively. The composite random variable $[X_1, X_3]$ assigns pairs of numbers to the outcomes in accordance with the following rule:

outcome	number pair	outcome	number pair
[HHH]	[3, −7]	[TTT]	[0, 1]
[HHT]	[2, 1]	[TTH]	[1, 1]
[HTH]	[2, 1]	[THT]	[1, 1]
[THH]	[2, 1]	[HTT]	[1, 1] ◁

Now we generalize the concept of probability distribution of a random variable to that of the *joint* probability distribution of a composite random variable.

Definition 7.7 Let $[X_1, X_2]$ be a composite random variable, and $[x_1, x_2]$ a pair of numbers representing outcomes of the game. The event $\{X_1 = x_1, X_2 = x_2\}$ is the intersection of the events $\{X_1 = x_1\}$ and $\{X_2 = x_2\}$ as defined in Definition 7.2; in other words, it is the set of outcomes represented by the pair $[x_1, x_2]$.

Let $[X_1, X_2]$ be the composite random variable in Example 7.4; the events $[X_1 = x_1, X_2 = x_2]$ are:

event	consists of outcomes
$\{X_1 = 0, X_2 = 0\}$	[TTT]
$\{X_1 = 1, X_2 = 1\}$	[TTH], [THT], [HTT]
$\{X_1 = 2, X_2 = 1\}$	[HTH]
$\{X_1 = 2, X_2 = 2\}$	[HHT], [THH]
$\{X_1 = 3, X_2 = 3\}$	[HHH]

Let $[X_1, X_3]$ be the composite random variable in Example 7.5; the events $[X_1 = x_1, X_3 = x_2]$ are:

event	consists of outcomes
$\{X_1 = 0, X_3 = 1\}$	[TTT]
$\{X_1 = 1, X_3 = 1\}$	[TTH], [THT], [HTT]
$\{X_1 = 2, X_3 = 1\}$	[HHT], [HTH], [THH]
$\{X_1 = 3, X_3 = -7\}$	[HHH]

Definition 7.8 For each value $[x_1, x_2]$ of the composite random variable $[X_1, X_2]$ let $\Pr(X_1 = x_1, X_2 = x_2)$ be the probability of the event $\{X_1 = x_1, X_2 = x_2\}$. The combined system of such values and corresponding probabilities is called the joint probability distribution of $[X_1, X_2]$.

The joint probability distribution of the composite random variable $[X_1, X_2]$ in Example 7.4 is

$$\Pr(X_1 = 0, X_2 = 0) = q^3,$$
$$\Pr(X_1 = 1, X_2 = 1) = 3pq^2,$$
$$\Pr(X_1 = 2, X_2 = 1) = p^2q,$$
$$\Pr(X_1 = 2, X_2 = 2) = 2p^2q,$$
$$\Pr(X_1 = 3, X_2 = 3) = p^3.$$

The joint probability distribution of the composite random variable $[X_1, X_3]$ in Example 7.5 is

$$\Pr(X_1 = 0, X_3 = 1) = q^3,$$
$$\Pr(X_1 = 1, X_3 = 1) = 3pq^2,$$
$$\Pr(X_1 = 2, X_3 = 1) = 3p^2q,$$
$$\Pr(X_1 = 3, X_3 = -7) = p^3.$$

The idea of composite random variables formed from two random variables is now generalized to *two or more*.

Definition 7.9 Let $X_1, \ldots, X_k$ be a set of random variables representing the outcomes of the game. The composite random variable $[X_1, \ldots, X_k]$ is a representation of the outcomes by multiplets of k numbers: with each outcome is associated the multiplet of k numbers assigned by $X_1, \ldots, X_k$, respectively.

Example 7.6 Let X_1, X_2, and X_3 be the random variables in Examples 7.1, 7.2, and 7.3, respectively. The composite random variable $[X_1, X_2, X_3]$ assigns triples of numbers to the outcomes of three tosses in accordance with the following rule:

outcome	triple	outcome	triple
[HHH]	[3, 3, −7]	[TTT]	[0, 0, 1]
[HHT]	[2, 2, 1]	[TTH]	[1, 1, 1]
[HTH]	[2, 1, 1]	[THT]	[1, 1, 1]
[THH]	[2, 2, 1]	[HTT]	[1, 1, 1]

◁

Definition 7.10 Let $[X_1, \ldots, X_k]$ be a composite random variable, and $[x_1, \ldots, x_k]$ a multiplet of k numbers representing outcomes of the game. The event $\{X_1 = x_1, \ldots, X_k = x_k\}$ is the intersection of the events $\{X_1 = x_1\}, \ldots, \{X_k = x_k\}$; in other words, it is the set of outcomes represented by the multiplet $[x_1, \ldots, x_k]$.

Let $[X_1, X_2, X_3]$ be the composite random variable defined in Example 7.6; the events $\{X_1 = x_1, X_2 = x_2, X_3 = x_3\}$ are:

event	consists of outcomes
$\{X_1 = 3, X_2 = 3, X_3 = -7\}$	[HHH]
$\{X_1 = 2, X_2 = 2, X_3 = 1\}$	[HHT], [THH]
$\{X_1 = 2, X_2 = 1, X_3 = 1\}$	[HTH]
$\{X_1 = 1, X_2 = 1, X_3 = 1\}$	[TTH], [THT], [HTT]
$\{X_1 = 0, X_2 = 0, X_3 = 1\}$	[TTT].

Definition 7.11 For each value $[x_1, \ldots, x_k]$ of the composite random variable $[X_1, \ldots, X_k]$ let $\Pr(X_1 = x_1, \ldots, X_k = x_k)$ be the probability of the event $\{X_1 = x_1, \ldots, X_k = x_k\}$. The combined system of such values and corresponding probabilities is called the joint probability distribution of $[X_1, \ldots, X_k]$.

The joint probability distribution of the composite random variable $[X_1, X_2, X_3]$ in Example 7.6 is

$$\Pr(X_1 = 3, X_2 = 3, X_3 = -7) = p^3,$$
$$\Pr(X_1 = 2, X_2 = 2, X_3 = 1) = 2p^2q,$$
$$\Pr(X_1 = 2, X_2 = 1, X_3 = 1) = p^2q,$$
$$\Pr(X_1 = 1, X_2 = 1, X_3 = 1) = 3pq^2,$$
$$\Pr(X_1 = 0, X_2 = 0, X_3 = 1) = q^3.$$

7.5 EXERCISES

1. Let Y and Z be the random variables in Exercises 1 and 2, respectively, of Section 7.1. Find the joint probability distribution of $[Y, Z]$.

2. Let Z and X be the random variables in Exercises 2 and 4, respectively, of Section 7.1. Find the joint probability distribution of $[Z, X]$.

3. A coin is tossed three times. Let X_1 be the Number of H's on the first two tosses and X_2 the Number of H's on the third toss (0 or 1). Find the joint probability distribution of $[X_1, X_2]$ when $p = .5, .6$.

4. Find the joint probability distribution of $[X, Y, Z]$, where X, Y, and Z are the random variables in Exercises 4, 1, and 2, respectively, of Section 7.1.

5. Let X_1, X_2, and X_3 be random variables representing Abel's net gain after two, four, and six tosses, respectively. Find the joint probability distribution of $[X_1, X_2, X_3]$ when $p = .5$.

6. Let X represent Abel's Largest Fortune in four tosses defined as the largest difference between the Numbers of H's and T's recorded up to the completion of four tosses. Let Y be similarly defined as Abel's Smallest Fortune in four tosses. Find the joint probability distribution of $[X, Y]$ when $p = .5$.

7. Let X be Abel's Largest Fortune in four tosses (cf. Exercise 6), and Z Abel's net gain in four tosses. Find the joint probability distribution of $[X, Z]$ when $p = .5$.

8. Repeat Exercise 7 for three tosses.

chapter 8

RANDOM TRIALS AND
GENERAL RANDOM VARIABLES

The theory developed in Chapters 2 through 7 has been exclusively about the game of tossing a coin several times. All the definitions and propositions about events, probabilities, random variables, etc., were given in this context. Now we shall abstract the fundamental properties of the coin-tossing game and define a *general* game of chance, extending the definitions and propositions above. Since much of the reasoning in the particular case of coin tossing can be transposed to the general case, we shall not repeat much of the former; however, the reader should refer to it.

We begin with two more special games of chance: the drawing of balls from an urn and the tossing of a die. Later we introduce the abstract game of chance.

8.1 DRAWING BALLS FROM AN URN

Suppose an urn contains several balls, indistinguishable to the touch. If someone reaches into the urn and blindly picks out one, the *actual* ball that is drawn cannot be predicted with certainty before the drawing: it is a *variable* which may be realized as any one of the balls. To each ball we attribute a numerical value, a *probability*, standing for the probability that

101

the particular ball is the one that is actually drawn. The assignment of probabilities to the balls determines the mathematical model of the game. We shall say that a ball is *selected at random* from the urn if each ball is attributed the same probability of being drawn: if there are N balls in the urn, each is given probability $1/N$. (The probabilities are conventionally supposed to have the sum 1.)

Let the balls be of two colors, for example, white and red. The color of the ball that is actually drawn is either white or red. When the selection of the ball is at *random*, the probability that the ball selected is white (or red) is defined as the proportion of white (or red) balls in the urn; for example, if an urn contains three white and two red balls, then the probability of drawing a white is equal to $\frac{3}{5}$. If there are balls of *several* colors in the urn, then the probability of selecting a ball of a given color is equal to the proportion of balls of that color present in the urn.

Now we define the random selection of several balls from an urn. Let it contain N distinguishable balls; suppose a group of s balls is selected from these. The system of outcomes of the selection consists of all unordered samples of s balls from the population N (Chapter 1); in accordance with Proposition 1.3, the number of such samples is $\binom{N}{s}$. The actual unordered sample selected is a variable assuming the form of any of these.

In what follows we shall always use the term "sample" to mean *unordered* sample.

Definition 8.1 A sample of s balls is said to be drawn at random from an urn containing N balls ($s \leq N$) if each sample has the same probability $1/\binom{N}{s}$; in this case it is said that s balls have been "sampled at random" from the urn. The sample selected is called the Sample (capital S).

Example 8.1 A sample of $s = 3$ balls is selected at random from an urn containing $N = 5$ balls, labeled A, B, C, D, and E, respectively. The probability of the selection of the particular sample (A, D, E) is $1/\binom{5}{3} = \frac{1}{10}$.

Each sample of $s = 4$ balls from an urn containing $N = 7$ balls has probability $1/\binom{7}{4} = \frac{1}{35}$ under random sampling.　　　◁

Definition 8.2 The probability that the Sample contains exactly k red balls is defined as the ratio

$$\frac{\text{number of samples containing exactly } k \text{ red balls}}{\text{number of samples}}.$$

Example 8.2 An urn contains $N = 5$ balls of which two are red and three are white. Two balls are sampled at random from the urn ($s = 2$). Let us find the probabilities that there are 0, 1, or 2 red balls, respectively, in the Sample. Let A, B, C, D, and E be the labels on the balls; suppose that A and B represent red ones. The ten samples are:

$$(A, B)$$
$$(A, C) \quad (A, D) \quad (A, E) \quad (B, C) \quad (B, D) \quad (B, E)$$
$$(C, D) \quad (C, E) \quad (D, E)$$

The sample in the first row in the above display is the one containing two red balls; hence, the probability that the Sample contains two red balls is $\frac{1}{10}$. The samples in the second row each contain exactly one red ball; hence, the probability that the Sample contains one red ball is $\frac{6}{10}$. The samples in the third row contain no red balls; hence, the probability that the Sample contains no red balls is $\frac{3}{10}$.   ◁

Proposition 8.2 below furnishes a formula for the calculation of the probability that the Sample contains exactly k red balls; the following is a preliminary.

Proposition 8.1 Let an urn contain r red balls and w white balls ($r + w = N$); let k be a nonnegative integer not greater than r or s. The number of samples of s balls containing exactly k red balls is equal to

$$\binom{r}{k} \binom{w}{s - k}. \tag{8.1}$$

$r = $ red color ball
$s = $ # of balls in sample

PROOF

1. The number of ways of forming a sample of k (red) balls from a population of r (red) balls is $\binom{r}{k}$.

Reason. The latter is the number of (unordered) samples of size k from a population r: apply Proposition 1.3 with r and k in place of N and s, respectively.

2. The number of ways of forming a sample of $s - k$ (white) balls from a population of w (white) balls is $\binom{w}{s-k}$.

Reason. This is analogous to Statement 1: replace k and r by $s - k$ and w, respectively.

3. The number of samples with exactly k red balls and $s - k$ white balls is given by formula (8.1).

Reason. Such a sample is obtained by combining a sample of k red balls with a sample of $s - k$ white balls. Statements 1 and 2 give the numbers of such samples: with each of $\binom{r}{k}$ samples of red balls, we combine one of $\binom{w}{s-k}$ samples of white balls; therefore, there are $\binom{r}{k}\binom{w}{s-k}$ samples with exactly k red and $s - k$ white balls.

4. The number of samples with exactly k red balls is given by formula (8.1).

Reason. This follows from Statement 3 and the fact that a sample of size s with k red balls necessarily contains $s - k$ white balls. ◀

Formula (8.1) permits an easy calculation of the probabilities sought in Example 8.2. The numbers of samples of size 2 containing 0, 1, and 2 red balls are

$$\binom{2}{0}\binom{3}{2} = 3, \qquad \binom{2}{1}\binom{3}{1} = 6, \qquad \binom{2}{2}\binom{3}{0} = 1,$$

respectively; hence, the probabilities are obtained by division by 10.

Example 8.3 An urn contains $r = 5$ red and $w = 4$ white balls. The numbers of samples of size $s = 3$ containing exactly 1, 2, and 3 red balls are

$$\binom{5}{1}\binom{4}{2} = 30, \qquad \binom{5}{2}\binom{4}{1} = 40, \qquad \binom{5}{3}\binom{4}{0} = 10,$$

respectively. (This example is continued below.)

> **Proposition 8.2** *An urn contains N balls of which r are red and w are white, with $r + w = N$. Suppose that s balls are sampled at random, and that k is a nonnegative integer not greater than either s or r. The probability that exactly k red balls are included in the Sample is*
>
> $$\binom{r}{k}\binom{w}{s-k} \Big/ \binom{N}{s}. \qquad\qquad (8.2)$$

PROOF. According to Proposition 1.1, the number of samples is given by the denominator in formula (8.2); by Proposition 8.1, the number of samples containing exactly k red balls is given by the numerator; and by Definition 8.2, the probability is equal to the given quotient. ◀

Example 8.3 (Continued). The probabilities that the Sample contains 1, 2, and 3 red balls are $\frac{30}{84}$, $\frac{40}{84}$, and $\frac{10}{84}$, respectively, because the total number of samples is $\binom{9}{3} = 84$.   ◁

8.1 EXERCISES

In each of the following, there are N balls in an urn, of which r are red and w are white; a sample of s is selected at random.

1. For $N = 7$, $r = 4$, $w = 3$, $s = 4$, find the probabilities that the Sample contains 1, 2, 3, 4 red balls, respectively.

2. For $N = 8$, $r = 5$, $w = 3$, $s = 3$, find the probabilities that the Sample contains 0, 1, 2, 3 white balls, respectively.

3. For $N = 6$, $r = 3$, $s = 2$, find the probabilities that the Sample contains 0, 1, 2 red balls, respectively.

4. For $N = 9$, $r = 7$, $w = 2$, $s = 3$, find the probabilities that the Sample contains 0, 1, 2, 3 red balls, respectively.

8.2 THE DICE GAME

When a six-sided die is tossed, it rolls and then comes to rest with one of its sides facing up. The six sides are labeled by numbers of black dots equal to 1, 2, 3, 4, 5, and 6, respectively. The actual number that turns up as a result of a toss cannot be predicted with certainty before the toss: it is a variable and assumes a value equal to one of the first six positive integers. This variable will be referred to as the Score (with capital S) on the toss of the die. We assign probabilities to each of the six sides: the probability of a particular side is the "probability that the Score takes on the value of the number of dots on the side"; for example, the probability assigned to the side with the number 3 is the "probability that the Score takes on the value 3." The die is said to be balanced if each of the six sides is assigned the same probability. We take each of the probabilities to be $\frac{1}{6}$, so that they have the sum 1.

    Suppose the die is tossed twice, and that the two tosses are done under "identical conditions": by this we mean that (a) the die is balanced on both tosses, and (b) the Score on one toss does not affect the Score on the other. These are assumptions on the *physical* and *mechanical* properties of the die and the conditions under which it is tossed; now we outline an appropriate *mathematical* model. The two tosses have exactly one of 36 outcomes; these are represented as the number pairs in the following table, where the first number in the bracket is an outcome of the first toss and the second number an outcome of the second:

$$
\begin{array}{cccccc}
[1\ 1] & [1\ 2] & [1\ 3] & [1\ 4] & [1\ 5] & [1\ 6] \\
[2\ 1] & [2\ 2] & [2\ 3] & [2\ 4] & [2\ 5] & [2\ 6] \\
[3\ 1] & [3\ 2] & [3\ 3] & [3\ 4] & [3\ 5] & [3\ 6] \\
[4\ 1] & [4\ 2] & [4\ 3] & [4\ 4] & [4\ 5] & [4\ 6] \\
[5\ 1] & [5\ 2] & [5\ 3] & [5\ 4] & [5\ 5] & [5\ 6] \\
[6\ 1] & [6\ 2] & [6\ 3] & [6\ 4] & [6\ 5] & [6\ 6]
\end{array}
\tag{8.3}
$$

We say that these 36 outcomes are "equally likely" and assign probability $\frac{1}{36}$ to each; for example, the probability that the Scores on the first and second tosses are 2 and 5, respectively, is equal to $\frac{1}{36}$.

A gambler is usually interested in the sum of the Scores turning up on the tosses. The possible sums for two tosses are 2, 3, . . . , 12. We define the probability of a particular sum as the sum of the probabilities of the pairs having the particular sum; for example, the probability that the sum of the two Scores is 3 is the sum of the probabilities for the pairs [2 1] and [1 2], namely, $\frac{1}{18}$; and, for example, the probability that the sum of the Scores is 7 is the sum of the probabilities of the six pairs [6 1], [5 2], [4 3], [3 4], [2 5], and [1 6]. Here is a table of the possible sums and their corresponding probabilities:

sum of Scores:	2	3	4	5	6	7	8	9	10	11	12	
probability:	$\frac{1}{36}$	$\frac{2}{36}$	$\frac{3}{36}$	$\frac{4}{36}$	$\frac{5}{36}$	$\frac{6}{36}$	$\frac{5}{36}$	$\frac{4}{36}$	$\frac{3}{36}$	$\frac{2}{36}$	$\frac{1}{36}$	(8.4)

Suppose the die is tossed *three* times under the following mechanical conditions: (a) the die is balanced on each toss, and (b) the Scores on the three tosses have no mutual effects; then, the following mathematical model is appropriate. The possible outcomes are representable as the set of 216 triples formed from the integers 1 through 6: [1 1 1], [2 1 1], [1 2 1], . . . , [6 6 5], [6 6 6]. We assign equal probabilities $\frac{1}{216}$ to each of these. The sum of the three Scores may be any one of the integers 3, 4, . . . , 18. By analogy with the case of two tosses, we define the probability of a particular sum as the sum of the probabilities of the triples having the particular sum; for example, the probability that the sum of Scores is 3 is the probability of the triple [1 1 1], namely, $\frac{1}{216}$; and the probability that the sum of Scores is 4 is the sum of the probabilities of the triples [2 1 1], [1 2 1],

[1 1 2], namely, $\frac{3}{216} = \frac{1}{72}$. There are six triples ([2 2 1], [2 1 2], [1 2 2], [1 1 3], [1 3 1], [3 1 1]) having the sum 5; hence, the probability that the sum of Scores is 5 is equal to $\frac{6}{216} = \frac{1}{36}$.

The principles for two and three tosses can be extended to any number n of tosses of the die. The outcomes of the tosses are multiplets of n integers, each being one of the first six integers. There are 6^n such multiplets and each is assigned probability 6^{-n}. The probability that the sum of Scores is equal to a particular number k is the sum of the probabilities of all the multiplets whose numbers have the sum k.

8.2 EXERCISES

1. Verify the probabilities in (8.4) by actual computation.

2. Construct a table of probabilities like (8.4) for the probabilities of the sum of Scores in three tosses. Is 10 or 11 more probable?

3. In five tosses of a die, what is the probability that the sum of the Scores is 8? 9?

4. In four tosses, what is the probability that the sum of the Scores is 7? 20?

5. In 10 tosses, what is the probability that the sum of the Scores is 12?

8.3 ABSTRACT RANDOM TRIALS

We have presented three different models of "games" or "trials" having the common property that the game has several *possible* outcomes, and that the *actual* outcome cannot be predicted with certainty but is determined by "chance"; these models are

a) the coin-tossing game;
b) the drawing of balls at random from an urn; and
c) the die-tossing game.

Now we construct an abstraction of a "game of chance." The three games described above, and many others, are concrete examples of this abstract game. We begin with the undefined concept of "system of outcomes"; then we introduce probabilities, events, and random variables, just as for the coin-tossing game.

⌈ **Definition 8.3** A *random trial* is a finite system of outcomes. Each is assigned a nonnegative number called its probability. The sum of the probabilities is taken to be 1. The probability assigned to a particular outcome is called the "probability of the outcome." ⌋

The coin game was our first example of a random trial: in a game of n tosses, the system of outcomes and their probabilities are given by Definitions 2.1 and 2.2.

The second example was the random sampling of s balls from an urn containing N: the system of outcomes is the system of all unordered samples of s balls, and their corresponding probabilities are all equal to $1/\binom{N}{s}$.

The third example was the dice game. When the die is tossed n times, the system of outcomes is the system of all multiplets of n integers, each being one of the integers $1, \ldots, 6$. There are 6^n such multiplets; each is assigned probability 6^{-n}.

The concepts of *event* for a random trial and *probability* of an event are the same as in the particular case of the coin-tossing game: an event is a set of outcomes, and the probability of an event is the sum of the probabilities of the outcomes forming it; furthermore, *all the definitions and propositions of Section 5.1 may be carried over word for word from the case of the coin game to that of the abstract game.* With the exception of Proposition 5.1, all the definitions and proofs in that section are entirely independent of the *particular form* of the coin game and are valid for the abstract game. Our purpose in casting the contents of that section in the terms of the coin game was to give the reader a concrete example to reason with. Though the *proof* of Proposition 5.1 is valid only in the particular case of the coin game, the truth of the proposition is, in the abstract case, assumed: the sum of the probabilities is 1 in Definition 8.3.

Now we show that the definitions of probabilities given in Sections 8.1 and 8.2 for the ball-sampling and dice games, respectively, are consistent with the definitions in the abstract case. Suppose an urn contains N balls of which some are red and some white. If s balls are sampled at random, the probability that the Sample contains k red balls is given in Definition 8.2 as the ratio of the number of samples with exactly k red balls to the number of samples. If we define the event "the Sample has k red balls" as the set of samples (outcomes) with k red balls, then its probability, as defined in the abstract case, is the sum of the probabilities of outcomes forming the event, and conforms to Definition 8.2. Suppose a die is tossed twice; and let us define the event "the sum of the Scores is x" as the set of all pairs (outcomes) in formula (8.3) having the sum x. The probability of this event, as abstractly defined, is the sum of the probabilities of the pairs forming the event, and conforms to the definition given in Section 8.2.

In the exercises that follow, the concepts in Section 5.1 are illustrated in the cases of the ball-sampling and dice games.

8.3 EXERCISES

1. An urn contains six balls labeled A, B, C, D, E, and F, of which A, B, and C are red and the rest are white. Two balls are sampled at random. Let M be the event "no red balls are in the Sample," and N the event "exactly one red ball is in the sample."

a) Enumerate the outcomes in the events M and N.
b) Describe, as in (a), the union and intersection of M and N, and their complements.
c) Find the probability of each event in (a) and (b).

2. An urn contains five balls labeled A, B, C, D, and E, of which A, B, and C are red and D and E white. Three balls are sampled at random from the urn. Let L, M, and N be the events

$$L = \text{"no red balls are in the Sample,"}$$
$$M = \text{"exactly one red ball is in the Sample,"}$$
$$N = \text{"at least two red balls are in the Sample."}$$

a) Enumerate the outcomes in each of the above events, and find their probabilities.
b) Let K be the union of L and M. Show that K is the complement of N.

3. A die is tossed twice. Let A, B, and C be the events

$$A = \text{"the sum of the Scores is at least 4,"}$$
$$B = \text{"the sum of the Scores is at most 10,"}$$
$$C = \text{"the sum of the Scores is 3."}$$

a) Enumerate the outcomes in each of the above events, and find their probabilities.
b) Describe, as in (a), the unions of A and B, of A and C, and of B and C.
c) What are the probabilities of the intersections of A and B, of A and C, and of A, B, and C?
d) Describe the complements of A and B.

4. A die is tossed three times. Let A and B be the events

$$A = \text{"the sum of the Scores is at most 4,"}$$
$$B = \text{"the sum of the Scores is at least 16."}$$

Find the probabilities of A and B, their union, intersection, and complements.

8.4 ABSTRACT RANDOM VARIABLES

The definition of random variable for an abstract random trial is literally the same as that for the coin game in Chapter 7: it is a numerical representation of the system of outcomes, a function which associates a real number with each outcome. Examples of random variables for the coin-tossing game are given in Examples 7.1, 7.2, and 7.3; now we give examples from the ball-sampling and dice games.

Example 8.4 An urn contains five balls labeled A, B, C, D, and E; A, B, and C are red and D and E are white. Two balls are sampled at random from the urn. Consider the function which associates numbers with the outcomes as follows:

outcome (sample)	number	outcome	number
(AB)	2	(BD)	1
(AC)	2	(BE)	1
(AD)	1	(CD)	1
(AE)	1	(CE)	1
(BC)	2	(DE)	0

This function is the random variable that we shall describe as "the number of red balls in the Sample." ◁

Example 8.5 Consider the urn model in the previous example, but disregard the color of the balls. Let us set up the following association:

outcome	number	outcome	number
(AB)	1	(BD)	0
(AC)	1	(BE)	1
(AD)	1	(CD)	0
(AE)	2	(CE)	1
(BC)	0	(DE)	1

The random variable may be described as "the number of balls in the Sample with vowel-letter labels." ◁

Example 8.6 A die is tossed twice. The function which associates the sum of the Scores with each outcome is a random variable called "the sum of the Scores."   ◁

Example 8.7 Consider the following random variable for two tosses of a die:

outcome	number	outcome	number	outcome	number
[1 1]	0	[3 1]	2	[5 1]	4
[1 2]	−1	[3 2]	1	[5 2]	3
[1 3]	−2	[3 3]	0	[5 3]	2
[1 4]	−3	[3 4]	−1	[5 4]	1
[1 5]	−4	[3 5]	−2	[5 5]	0
[1 6]	−5	[3 6]	−3	[5 6]	−1
[2 1]	1	[4 1]	3	[6 1]	5
[2 2]	0	[4 2]	2	[6 2]	4
[2 3]	−1	[4 3]	1	[6 3]	3
[2 4]	−2	[4 4]	0	[6 4]	2
[2 5]	−3	[4 5]	−1	[6 5]	1
[2 6]	−4	[4 6]	−2	[6 6]	0

This random variable is "the difference between the first and second Scores."   ◁

All of the formal definitions and propositions in Section 7.1 are transposed word for word to abstract random variables:

The set of values of a random variable is the set of numbers in the corresponding numerical representation of the outcomes.

For a number x in the set of values of a random variable X, the set of outcomes numerically represented by X is called the event $\{X = x\}$.

The probability distribution of X is the system of values x and corresponding probabilities $\Pr(X = x)$. The expected value $E(X)$ is the same sum as in Definition 7.4, and the variance is as in Definition 7.5.

Let X_1 and X_2 be random variables for a random trial. The composite random variable $[X_1, X_2]$ is defined just as in Definition 7.6, and the joint probability distribution is as in Definition 7.8. A composite random variable formed from two or more random variables on a random trial is defined as in Definition 7.9, and the joint probability distribution is as in Definition 7.11.

Let X_1, X_2, Y_1, and Y_2 be the random variables in Examples 8.4, 8.5, 8.6, and 8.7, respectively; their probability distributions are:

x:	0	1	2
$\Pr(X_1 = x)$:	.1	.6	.3

x:	0	1	2
$\Pr(X_2 = x)$:	.3	.6	.1

[The probability distribution of Y_1 is given in (8.4).]

y:	-5	-4	-3	-2	-1	0	1	2	3	4	5
$\Pr(Y_2 = y)$:	$\frac{1}{36}$	$\frac{2}{36}$	$\frac{3}{36}$	$\frac{4}{36}$	$\frac{5}{36}$	$\frac{6}{36}$	$\frac{5}{36}$	$\frac{4}{36}$	$\frac{3}{36}$	$\frac{2}{36}$	$\frac{1}{36}$

The expected values are:

$$E(X_1) = 0(.1) + 1(.6) + 2(.3) = 1.2,$$
$$E(X_2) = 0(.3) + 1(.6) + 2(.1) = .8,$$
$$E(Y_1) = 2\tfrac{1}{36} + 3\tfrac{2}{36} + 4\tfrac{3}{36} + \cdots + 12\tfrac{1}{36} = 7,$$
$$E(Y_2) = -5\tfrac{1}{36} - 4\tfrac{1}{36} - \cdots + 0\tfrac{6}{36} + \cdots + 4\tfrac{1}{36} + 5\tfrac{1}{36} = 0.$$

The variances of X_1 and X_2 are:

$$\mathrm{Var}(X_1) = (1.2)^2(.1) + (.2)^2(.6) + (.8)^2(.3) = .360,$$
$$\mathrm{Var}(X_2) = (.8)^2(.3) + (.2)^2(.6) + (1.2)^2(.1) = .360.$$

We leave it as an exercise for the reader to show that the variances of Y_1 and Y_2 are both equal to $\frac{35}{6}$.

The joint probability distribution of $[X_1, X_2]$ is computed:

event	consists of	probability
$\{X_1 = 0, X_2 = 1\}$	(DE)	.1
$\{X_1 = 1, X_2 = 0\}$	(BD), (CD)	.2
$\{X_1 = 1, X_2 = 1\}$	(AD), (BE), (CE)	.3
$\{X_1 = 1, X_2 = 2\}$	(AE)	.1
$\{X_1 = 2, X_2 = 0\}$	(BC)	.1
$\{X_1 = 2, X_2 = 1\}$	(AB), (AC)	.2

The joint probability distribution of $[Y_1, Y_2]$ is the system of probabilities of 36 events of the form $\{Y_1 = y_1, Y_2 = y_2\}$, each consisting of a single outcome from among the 36 possible outcomes of the two tosses; for example, the event $\{Y_1 = 5, Y_2 = -1\}$ consists of the outcome [2 3]. It follows that each of the probabilities is $\frac{1}{36}$.

8.4 EXERCISES

1. Show that the variances of the random variables in Examples 8.6 and 8.7 are $\frac{35}{6}$.

2. An urn contains six balls labeled by the first six integers. The balls numbered 1, 2 are red and the others are white. Two balls are sampled at random from the urn. Let X_1 and X_2 be random variables representing the number of red balls in the Sample, and the number of even-numbered balls in the Sample, respectively. Find the probability distributions of X_1 and X_2, the expected values, variances, and their joint probability distribution.

3. A die is tossed twice. Let X_1 and X_2 be the Score on the first toss, and the sum of the Scores on the two tosses, respectively. Find the joint probability distribution of X_1 and X_2.

4. Let X be the *square* of the Score on a single toss of a die. Find the probability distribution, expected value, and variance of X.

5. Consider the abstract random trial with the outcomes $t_1, t_2, \ldots, t_8$, and corresponding probabilities .1, .1, .2, .2, .1, \ldots, .1, respectively. Let the random variables X_1, X_2, and X_3 numerically represent the outcomes in the following way:

outcome	values assigned by		
	X_1	X_2	X_3
t_1	1	1	1
t_2	2	2	2
t_3	1	3	3
t_4	2	1	4
t_5	1	2	1
t_6	2	3	2
t_7	1	1	3
t_8	2	2	4

Find the probability distributions of X_1, X_2, and X_3; and the joint probability distributions of $[X_1, X_2]$, $[X_1, X_3]$, $[X_2, X_3]$, and $[X_1, X_2, X_3]$.

6. A balanced coin is tossed three times. Let X be the difference between the Numbers of H's and T's. Find $E(X)$ and $\text{Var}(X)$.

7. Cain and Abel play a five-toss game. Let the random variable X be the number of times in the course of the five tosses that Abel's Fortune is 0. Find $E(X)$ when $p = .5$.

8. (Refer to Exercise 7.) Let Y be the number of times in the course of the five tosses that Abel's Fortune is positive. Find $E(Y)$.

9. A die is tossed twice. Let X be the *average* of the Scores. Find the probability distribution of X.

10. Find the expected values and variances of the random variables X_1, X_2, and X_3 in Exercise 5.

chapter 9

INDEPENDENT RANDOM VARIABLES, EMPIRICAL FREQUENCIES, AND THE CLUSTERING PRINCIPLE

Much of the interest in probability is a consequence of its importance in the interpretation of mass numerical data. (We call such data *observed*, or *empirical*.) They can often be mathematically described as the collection of the actual values of *independent* random variables with a common probability distribution. In this chapter we examine the relation between the observed data and the underlying common probability distribution. The main result is the Clustering Principle, a fundamental theorem of mathematical statistics; its proof is based on the concept of independence of random variables (Section 9.2) and the Law of Large Numbers.

9.1 THE IMPORTANCE OF RANDOM VARIABLES

In many random trials one is interested more in some numerical representation of the outcomes than in the individual outcomes. Consider the coin-tossing game: in all the applications of this model in Chapter 2, we were interested not in the particular Outcome of n tosses, that is, not in the particular multiplet of H's and T's, but rather in the Number of H's. In the second example of a random trial, the urn with balls of two colors, the number

114

of balls of a particular color in the Sample was of special interest. In the third example, the die-tossing game, the sum of the Scores was of special interest. In each case our attention moved from the *original* system of outcomes and their probabilities to the *set of values* of the random variable and its *probability distribution*.

There are many phenomena in the natural and social world which can be mathematically described as random trials. We have already discussed some which are describable as a coin-tossing game; however, many processes of interest are too complicated to be described so simply. The daily weather is an example: its description is a complex of time- and place-dependent measurements which are not completely predictable. If today's weather is considered as a random trial, then the system of outcomes consists of the entire description of possible temperature, air pressure, wind, humidity, etc., measurements at each of many different places and times. Fortunately, most of the public is interested, not in the particular outcomes of such a random trial, but only in a few numerical representations—random variables —of the weather conditions, such as maximum and minimum temperatures and amount of precipitation. Let X be the random variable representing maximum daily temperature (Fahrenheit); if it is measured to the nearest degree, its set of values is the set of integers. The probability distribution of X depends on the time of year and the place; for example, the probability $Pr(X = 90)$ will be different in New York in July and in Chicago in January.

A primary problem of statistics is the estimation of probability distributions of random variables. The income of the farmer whose profit is greatly influenced by the weather is like the fortune of a gambler whose gains are determined by a random variable on a random trial; hence, the probability distributions of random variables associated with weather conditions are important to the farmer.

9.2 INDEPENDENT RANDOM VARIABLES

The concept of independent events was introduced in Chapter 5 in the framework of the coin-tossing game; now we introduce the notion of independent random variables.

Definition 9.1 Let $[X_1, X_2]$ be a composite random variable on a random trial; X_1 and X_2 are called independent if, for each value $[x_1, x_2]$ of $[X_1, X_2]$, the probability $Pr(X_1 = x_1, X_2 = x_2)$ is equal to the product of the probabilities $Pr(X_1 = x_1)Pr(X_2 = x_2)$. More generally, if $X_1, \ldots, X_k$ are random variables on a common random trial, they are called independent if each probability $Pr(X_1 = x_1, \ldots, X_k = x_k)$ of the joint probability distribution is equal to the corresponding product of probabilities $\Pr(X_1 = x_1) \cdots \Pr(X_k = x_k)$.

Example 9.1 A coin is tossed three times. Let X_1 be the Number of H's on the first two tosses, and X_2 the Number of H's on the third toss (0 or 1). The eight outcomes are represented by $[X_1, X_2]$, whose joint probability distribution is

$$\Pr(X_1 = 2, X_2 = 1) = p^2 \cdot p = \Pr(X_1 = 2)\Pr(X_2 = 1),$$
$$\Pr(X_1 = 2, X_2 = 0) = p^2 \cdot q = \Pr(X_1 = 2)\Pr(X_2 = 0),$$
$$\Pr(X_1 = 1, X_2 = 1) = 2pq \cdot p = \Pr(X_1 = 1)\Pr(X_2 = 1),$$
$$\Pr(X_1 = 0, X_2 = 1) = q^2 \cdot p = \Pr(X_1 = 0)\Pr(X_2 = 1),$$
$$\Pr(X_1 = 1, X_2 = 0) = 2qp \cdot q = \Pr(X_1 = 1)\Pr(X_2 = 0),$$
$$\Pr(X_1 = 0, X_2 = 0) = q^2 \cdot q = \Pr(X_1 = 0)\Pr(X_2 = 0).$$

The condition of Definition 9.1 is satisfied, so that X_1 and X_2 are independent.   ◁

Example 9.2 A die is tossed twice. Let X_1 and X_2 be the Scores on the two respective tosses. They are independent because each outcome in formula (8.3) has probability $\frac{1}{36}$, which is the product of the probabilities of the individual outcomes of X_1 and X_2; for example,

$$\Pr(X_1 = 4, X_2 = 5) = \tfrac{1}{36} = \tfrac{1}{6} \cdot \tfrac{1}{6} = \Pr(X_1 = 4)\Pr(X_2 = 5).   ◁$$

More generally, if $X_1, \ldots, X_n$ are the Scores on n successive tosses, then they are independent random variables.

Example 9.3 Here is an example of a pair of random variables which are *not* independent: X_1 and X_2 in Examples 8.4 and 8.5, respectively. The probability distributions of X_1 and X_2 and their joint probability distribution were computed in Section 8.4. Compare the probability

$$\Pr(X_1 = 0, X_2 = 1)$$

with the product of probabilities $\Pr(X_1 = 0)\Pr(X_2 = 1)$: the former is .1 and the latter is $(.1)(.6) = .06$; thus, one of the defining equations of independence is not satisfied; therefore, X_1 and X_2 are not independent.   ◁

The property of independence is one of a *family* of random variables; it is not a property of *individual* random variables. Here is a "consistency" trait of such a family.

Proposition 9.1 *Any subfamily of a family having the independence property also has that property: if $X_1, \ldots, X_n$ are independent random variables, so are any of them forming a subfamily, for example, $X_1, \ldots, X_k, k < n$.*

PROOF. We shall give the proof for just a family of three random variables; the proof for the general case proceeds in the same way but requires more notation. We shall now prove: If X, Y, and Z are independent random variables, then so are X and Y. Here x and y represent values of X and Y, respectively; $z_1, z_2, \ldots$ form the set of values of Z.

1. $\Pr(X = x, Y = y)$ is equal to the sum of the probabilities

$$\Pr(X = x, Y = y, Z = z_1), \qquad \Pr(X = x, Y = y, Z = z_2), \ldots .$$

Reason. The events $\{Z = z_1, Z = z_2\}, \ldots$ are disjoint and their union is the sure event (Proposition 5.5).

2. $\Pr(X = x, Y = y, Z = z) = \Pr(X = x)\Pr(Y = y)\Pr(Z = z)$ holds for every $z = z_1, z_2, \ldots .$

Reason. X, Y, and Z are independent.

3. The sum of probabilities in Statement 1 is equal to $\Pr(X = x)\Pr(Y = y)$.

Reason. By Statement 2 this sum is equal to the sum

$$\Pr(X = x)\Pr(Y = y)\Pr(Z = z_1) + \Pr(X = x)\Pr(Y = y)\Pr(Z = z_2) + \cdots .$$

When the common factors $\Pr(X = x)\Pr(Y = y)$ are removed from each term, the sum of the remaining factors is $\Pr(Z = z_1) + \Pr(Z = z_2) + \cdots$, which is equal to 1 (Proposition 7.1).

The assertion of the proposition is implied by Statements 1 and 3. ◀

We conclude this section with some remarks on the empirical significance of independence of random variables. In our theoretical model of the coin-tossing game, the Numbers of H's appearing on different tosses (0 or 1) are independent random variables; indeed, let $X_1, X_2, \ldots$ be the Numbers (0 or 1) of H's on the first, second, ... tosses, respectively; then

$$\Pr(X_1 = 1, X_2 = 0, \ldots) = pq \cdots = \Pr(X_1 = 1) \cdot \Pr(X_2 = 0) \cdots ,$$

so that the X's are independent. The Scores on successive tosses of a die are also independent random variables (Example 9.2). Experience has made known that our theoretical models give correct and useful descriptions of many real phenomena; however, not all or any theoretical models are appropriate for any given phenomenon. It is the function of statistics and science to determine the suitability of models. Let us illustrate this in the study of the recovery of a patient from disease (Section 2.2). The concept of independence in the coin-tossing game is the "mathematical expression" of the empirical fact that the outcomes on successive tosses have no "mutual effects"; thus, this model is appropriate for the study of disease only if the recoveries of patients are without mutual effects among them. Only medical statistics—not mathematics—can verify or refute this. The same is true for any multiple measurement of a phenomenon represented as a composite random variable: If the component measurements are generated by occurrences that have no mutual effects, then the mathematical assumption of the independence of the component random variables is appropriate.

9.2 EXERCISES

1. Determine whether or not the random variables Y_1 and Y_2 in Examples 8.6 and 8.7, respectively, are independent.

2. A coin is tossed six times. Let X_1, X_2, and X_3 be the Number of H's on the first two tosses, the Number of T's on the second two tosses, and the Number of H's on the last two tosses, respectively. Prove that X_1, X_2, and X_3 are independent random variables.

3. An urn contains eight balls labeled 1, . . . , 8. The first four balls are red, and the last four are white. Two balls are selected at random from the urn. Put

$$X_1 = \text{number of white balls in the Sample,}$$
$$X_2 = \text{number of even-numbered balls in the Sample,}$$
$$X_3 = \text{number of balls in the Sample whose label number}$$
$$\text{is less than 6.}$$

Find the joint probability distributions of each of these composite random variables: $[X_1, X_2]$, $[X_1, X_3]$, $[X_2, X_3]$, $[X_1, X_2, X_3]$. Do any of these consist of independent random variables?

4. A die is tossed twice. Let X_1 be the Score on the first toss, and X_2 the square of the Score on the second toss. Find the joint probability distribution of X_1 and X_2. Are they independent?

5. Prove this version of Proposition 9.1: Any three out of four independent random variables are independent.

9.3 STATISTICAL ESTIMATION: THE FARMERS' ALMANAC

Consider the farmer who wants to know the probability distribution of the maximum daily temperature for the month of June in the district containing his farm. The probability distribution can be estimated by means of an *empirical frequency distribution* of recorded maximum temperatures, as follows.

For a given day—for example, June 1—the farmer finds the maximum temperatures recorded for that day on each preceding year for the last several years. (Suppose that this information is available to the farmer through almanacs.) Assume, for the purpose of illustration, that he has such records for the last 75 years. For each degree level, he records the *number* of years in which June 1 had that particular maximum temperature; then, the *proportion* of such years (the number of years divided by 75) is called the frequency of the given maximum temperature over the 75-year period. The empirical frequency distribution is a "good" estimate of the probability distribution of the random variable X representing the June 1 maximum temperature. Later we shall explain the meaning of "good" estimate and prove the above statement using the concepts of independent random variables and the Law of Large Numbers.

Example 9.4 Here is a list of the daily maximum temperatures for June 1 of the years 1875–1949 for Bismarck, North Dakota (U.S. Department of Commerce, Environmental Science Services Administration):

59	80	79	60	67
46	64	71	76	72
50	70	81	74	69
66	63	64	88	76
66	84	69	72	66
67	62	62	67	87
78	73	81	63	63
67	57	74	55	72
71	69	77	68	74
72	75	75	67	90
76	67	77	70	47
67	82	59	81	53
70	75	68	73	64
67	69	64	92	90
79	68	58	90	71

We arrange these in a frequency distribution, indicating the number of days for each degree:

temperature:	46	47	48	49	50	51	52	53
number of days:	1	1	–	–	1	–	–	1
frequency:	.013	.013	–	–	.013	–	–	.013

	54	55	56	57	58	59	60	61
	–	1	–	1	1	2	1	–
	–	.013	–	.013	.013	.027	.013	–

	62	63	64	65	66	67	68	69
	2	3	4	–	3	8	3	4
	.027	.040	.053	–	.040	.107	.040	.053

	70	71	72	73	74	75	76	77
	3	3	4	2	3	3	3	2
	.040	.040	.053	.027	.040	.040	.040	.027

	78	79	80	81	82	83	84	85
	1	2	1	3	1	–	1	–
	.013	.027	.013	.040	.013	–	.013	–

	86	87	88	89	90	91	92
	–	1	1	–	3	–	1
	–	.013	.013	–	.040	–	.013

This can be used to estimate the probability that the maximum temperature falls in a given temperature range; for example, the probability that the maximum temperature exceeds 69°. The number of days in which the temperature exceeded this is 38; therefore, the estimated probability of a maximum temperature exceeding 69° is $\frac{38}{75} = .507$. ◁

9.4 STATISTICAL CONFIRMATION OF HYPOTHETICAL PROBABILITY DISTRIBUTIONS

Suppose a die is repeatedly tossed and the Scores are recorded until a total of n tosses have been completed. Let $S_1, \ldots, S_6$ stand for the number of Scores equal to 1, ..., 6, respectively; for example, if the die is tossed 10 times ($n = 10$) and the Scores 3, 4, 1, 4, 6, 2, 6, 3, 5, and 6 are observed, then $S_1 = 1$, $S_2 = 1$, $S_3 = 2$, $S_4 = 2$, $S_5 = 1$, $S_6 = 3$. The quantities $S_1, \ldots, S_6$ are random variables on the random trial of n tosses of the die. Put $f_1 = S_1/n, \ldots, f_6 = S_6/n$; these are the relative frequences of $1, \ldots, 6$ in n tosses. If the number of tosses is very large, these frequencies will, with "high" probability, be very "close" to the theoretical probability $\frac{1}{6}$. This phenomenon is called the "clustering" of the observed frequencies about the theoretical probabilities. It can be empirically verified; the interested reader should try it himself.

One consequence of the clustering phenomenon is a law of large numbers for die tossing: the *average* of many Scores is, with high probability, close to the expected value of the Score on a single toss. This expected value is equal to $3\frac{1}{2}$; indeed, it is

$$1 \cdot \tfrac{1}{6} + 2 \cdot \tfrac{1}{6} + 3 \cdot \tfrac{1}{6} + 4 \cdot \tfrac{1}{6} + 5 \cdot \tfrac{1}{6} + 6 \cdot \tfrac{1}{6} = 3\tfrac{1}{2}.$$

The average of several Scores may be expressed as

$$1 \cdot f_1 + 2 \cdot f_2 + 3 \cdot f_3 + 4 \cdot f_4 + 5 \cdot f_5 + 6 \cdot f_6;$$

the latter sum is close to the former when $f_1, \ldots, f_6$ are all close to $\frac{1}{6}$. This illustrates the significance of the concept of expected value in this case. The General Law of Large Numbers will be derived in Chapter 12.

9.5 THE CLUSTERING PRINCIPLE

The assertions made in the farmer's estimation problem in Section 9.3 and in the die problem in Section 9.4 will now be put in an abstract form and deduced from previous results.

The Clustering Principle. *Let $X_1, \ldots, X_n$ be independent random variables on a random trial, having a common probability distribution. Let $x_1, \ldots, x_k$ be the set of values of each random variable. Let $f_1, \ldots, f_k$ be the random variables representing the relative frequencies of $x_1, \ldots, x_k$, respectively; for example, for a given outcome, f_1 is the ratio*

$$f_1 = \frac{1}{n} \cdot \text{number of events } \{X_1 = x_1\}, \ldots, \{X_n = x_1\}$$

containing the outcome.

For an arbitrary fixed positive number d, let P_n be the probability that the frequencies $f_1, \ldots, f_k$ are all within d units of the theoretical probabilities $\Pr(X_1 = x_1), \ldots, \Pr(X_1 = x_k)$, respectively.

The conclusion is: the limit of the sequence $P_1, P_2, \ldots, P_n, \ldots$ is 1; or, in other words, if the number of random variables is very large, the *probability is nearly* 1 *that the empirical frequencies are all close to the theoretical probabilities.*

The proof of the Clustering Principle is long and will be given in several steps. Before starting the proof, we show that the illustrations in Sections 9.3 and 9.4 are typical applications of the Principle. In the maximum temperature problem, the set of all possible weather conditions on every June 1 for the past 75 years is the random trial. The random variables $X_1, X_2, \ldots,$ X_{75} are the maximum temperatures on the successive first days of June. The assumption that the X's are independent is equivalent to the (meteorological) assumption that the maximum temperature on June 1 of one year is without influence on that of other years. The assumption that the X's have a common probability distribution is equivalent to the (meteorological) assumption that the long-term weather conditions are unchanged over the past 75 years. In this application the Clustering Principle implies that the observed frequencies of maximum temperatures are, with probability nearly 1, all close to the theoretical probabilities of maximum temperatures. In the tossing of the die (Section 9.4), the random trial is the system of all outcomes of n tosses of the die (Section 8.2). The random variables $X_1, \ldots, X_n$ are the Scores on the successive tosses. The assumption that the X's are independent with a common probability distribution ($\frac{1}{6}$ for each outcome) is just the assumption that the successive tosses have no mutual effects and that the die is not physically altered during the tossings. The Clustering Principle implies that the frequencies of $1, \ldots, 6$ are, with high probability, all close to $\frac{1}{6}$.

The proof begins. For a value x of the random variables $X_1, \ldots, X_n$ (x is one of the numbers $x_1, \ldots, x_k$), let S be the *numerator* in the ratio defining the frequency f of that value: for a particular outcome of the random trial, let S be the *number* of events $\{X_1 = x\}, \ldots, \{X_k = x\}$ containing the outcome. (In the data in the farmer's almanac problem, S assigns the number 2 to the value $x = 79$.) The first point in the proof is:

Proposition 9.2 *The random variable S has the probability distribution of the Number of H's in n tosses of a coin with $p = \Pr(X_1 = x)$.*

PROOF. With each of the random variables $X_1, \ldots, X_n$, let us associate the toss of a coin. For each outcome of the random trial, we observe whether or not it belongs to the events $\{X_1 = x\}, \ldots, \{X_n = x\}$, respectively: if the outcome belongs to $\{X_1 = x\}$, then we say that the first toss resulted in H; and if the outcome does not belong to $\{X_1 = x\}$, then we say that the first toss resulted in T. This identification is made for each of the random variables: if the outcome belongs to $\{X_j = x\}$, then we say that the jth toss of the coin yielded H; if not, then we say that T appeared on the jth toss; $j = 1, \ldots, n$. In this way, each outcome of the original random trial is associated with an outcome of the game of n tosses of a coin; for example, an outcome belonging to the intersection of the events: $\{X_1 = x\}$, $\{X_2 \neq x\}$, $\ldots$, $\{X_n = x\}$ is associated with the outcome [HT, $\ldots$, H] of the coin game. (Here we have used $\{X_2 \neq x\}$ to represent the complement of the event $\{X_2 = x\}$.) By its definition, the random variable S represents the Number of H's in this coin game; thus, in order to complete the proof of the proposition, we have to show that the *probabilities* assigned to the outcomes of *this* coin game are the same as in Definition 2.2, with p as asserted in the proposition and $q = 1 - p$; for example, we have to show that

$$\Pr\{X_1 = x, X_2 \neq x, \ldots, X_n = x\} = pq \cdots p.$$

The independence of the X's and the invariance of the independence property under the taking of complements of events (Proposition 5.8) are essential to the following proof.

1. $\Pr(X_1 = x, \ldots, X_n = x) = p \cdots p$ (n factors).

Reason. Independence of the X's, identity of their probability distributions, and the given definition of p.

2. The events $\{X_1 = x\}, \ldots, \{X_n = x\}$ are independent.

Reason. We have to show: The probability of the intersection of any number of them is equal to the product of the corresponding probabilities. Consider the intersection of the first k of them; this is a typical example. By virtue of Proposition 9.1, $X_1, \ldots, X_k$ are independent; thus, by Definition 9.1, the probability of the intersection of the k events is the product of corresponding probabilities.

3. The family of events obtained from the family $\{X_1 = x\}, \ldots, \{X_n = x\}$ by interchanging any of them with the corresponding complement $\{X \neq x\}$ is also a family of independent events.

Reason. Proposition 5.8.

4. The equation in Statement 1 continues to hold when any of the signs "$=$" in the event $X_1 = x, \ldots, X_n = x$ are changed to "$\neq$" and the corresponding factors p are changed to q.

Reason. Statement 3.

The last statement completes the proof of Proposition 9.2. ◄

The second step in the proof of the Clustering Principle is a direct application of an estimate used in the proof of the Law of Large Numbers:

Proposition 9.3 *For any value x and arbitrary positive number d, the probability that the frequency f = S/n is within d units of the theoretical probability is greater than or equal to*

$$1 - \frac{1}{4nd^2}.$$ (9.1)

PROOF. The probability that f is within d units of $p = \Pr(X = x)$ is at least equal to

$$1 - \frac{pq}{nd^2}$$

because of

a) the identification of f with the ratio of the Number of H's to the number of tosses (Proposition 9.2), and

b) the fundamental inequality in the proof of the Law of Large Numbers (Proposition 3.4).

The assertion of our proposition now follows from the inequality $pq \leq \frac{1}{4}$ (Proposition 3.3).   ◄

The proof of the Clustering Principle is not yet complete: though Proposition 9.3 states that the frequency of a *particular* value x tends, with high probability, to be near the theoretical probability $\Pr(X_1 = x)$, the Clustering Principle claims that the frequencies of all values $x_1, \ldots, x_k$ are *simultaneously* close to the respective probabilities $\Pr(X_1 = x_1), \ldots, \Pr(X_1 = x_k)$. Here is the next step in the proof.

Proposition 9.4 *Let $A_1, \ldots, A_k$ be events each having probability at least equal to $1 - c$, where c is some number between 0 and 1; then, the probability of the intersection of $A_1, \ldots, A_k$ is at least equal to $1 - kc$.*

PROOF

1. The probability of the complement of each of the events $A_1, \ldots, A_k$ is at most equal to c.

Reason. Proposition 5.3.

2. The intersection of $A_1, \ldots, A_k$ is obtained by starting with the whole set of outcomes and successively eliminating all outcomes in the complements of $A_1, \ldots, A_k$, respectively.

Reason. Definition 5.2.

3. The assertion of the proposition is valid.

Reason. The probability of the whole set of outcomes is 1. When the outcomes in the complements of $A_1, \ldots, A_k$, respectively, are successively eliminated, the corresponding probability is reduced by at most c at each step; hence, after k steps it is reduced by at most kc.   ◀

Proposition 9.4 implies: if each of the events $A_1, \ldots, A_k$ has probability "arbitrarily close" to 1, so does their intersection. We apply this to events $A_1, \ldots, A_k$ defined as:

$$A_1 = \text{``}f_1 \text{ is within } d \text{ units of } \Pr(X_1 = x_1)\text{,''}$$
$$\vdots \tag{9.2}$$
$$A_k = \text{``}f_k \text{ is within } d \text{ units of } \Pr(X_1 = x_k)\text{.''}$$

Proposition 9.5 *Let P_n be the probability of the intersection of the events $A_1, \ldots, A_k$ in formula (9.2); then P_n is at least equal to*

$$1 - \frac{k}{4nd^2} ;$$

thus the limit of the sequence $P_1, P_2, \ldots, P_n, \ldots$ is 1.

PROOF. Proposition 9.3 implies that the probability of each of the events A is at least equal to the quantity in formula (9.1); thus, Proposition 9.4 implies that the intersection has probability at least equal to that asserted above (put $c = 1/4nd^2$).

This completes the proof of the Clustering Principle.   ◀

9.5 EXERCISES

1. Consider 25 independent random variables with a common probability distribution and set of values $x_1 = 1$, $x_2 = 2$, and $x_3 = 3$; suppose the 25 observed values are:

$$1\ 1\ 2\ 1\ 2\ 3\ 2\ 2\ 1\ 1\ 2\ 1\ 1$$
$$2\ 1\ 1\ 2\ 1\ 3\ 1\ 1\ 2\ 1\ 3\ 3$$

Find f_1, f_2, and f_3.

2. Put three red and four black checkers in a box. Perform this experiment 25 times: draw two checkers at random from the box, replacing the drawn pair after each selection. Does the empirical frequency distribution of the number of black balls in the Sample approximately coincide with the theoretical probability distribution?

3. Toss a die 25 times and note the frequency distribution of the Scores; also toss a pair of dice 50 times and compare the frequencies of the sums of the two Scores to the theoretical probabilities.

4. Chart the derivation of the Clustering Principle from basic definitions; list the supporting propositions and the network of implications.

5. Composite random variables $[X_1, Y_1], \ldots, [X_n, Y_n]$ are called independent if

$$\Pr(X_1 = x_1, Y_1 = y_1, \ldots, X_n = x_n, Y_n = y_n)$$
$$= \Pr(X_1 = x_1, Y_1 = y_1) \cdots \Pr(X_n = x_n, Y_n = y_n)$$

for all values x and y. Extend the statement of the Clustering Principle to the frequencies and probabilities of independent composite random variables. Show that the proof is the same as for (ordinary) independent random variables.

6. At the end of Section 9.2 it was stated that the appropriateness of the independence assumption in any given real problem has to be justified by statistical analysis. What is the role of the Clustering Principle in such analysis? In particular, how can the Principle be used to justify the independence assumption for the Outcomes of two successive tosses of a real coin (cf. Exercise 5)?

9.6 APPLICATION TO AN EMPIRICAL "PROOF" OF THE NORMAL APPROXIMATION THEOREM

Suppose a balanced coin is tossed twice. The Number of H's has the set of values 0, 1, 2 to which are assigned probabilities $\frac{1}{4}, \frac{1}{2}, \frac{1}{4}$, respectively. If the pairs of tosses are repeated many times and the frequencies of the numbers of pairs with 0, 1, and 2 H's are recorded, then, with high probability, the frequencies will be very close to the respective assigned probabilities. In this way one can empirically check the theoretical model of the binomial distribution for $n = 2$, $p = \frac{1}{2}$; therefore, for the same reason, we can experimentally verify the normal approximation to the binomial distribution in Chapter 4.

Let us present the verification in the particular case $n = 20$, $p = \frac{1}{2}$. Toss the coin 20 times and let X_1 be the Number of H's; X_1 has the set of values 0, 1, ..., 20. Repeat the 20 tosses, and let X_2 be the Number of H's. Note that X_2 has the same probability distribution as X_1. The two random variables are _independent_: the events $\{X_1 = k\}$ and $\{X_2 = j\}$ (where k and j are arbitrary integers between 0 and 20, inclusive) are determined by different sets of tosses (Definition 5.7); hence, they are independent (Independence Theorem); thus, the probability of their intersection is the product of their respective probabilities (Definition 5.8), and so the random variables are independent (Definition 9.1). Now let $X_1, X_2, X_3, \ldots, X_m$ be the random variables obtained by performing the 20 tosses m times; by the foregoing argument, they are independent and have a common (binomial) probability distribution. If m is large, the observed frequencies of $0, 1, \ldots, 20$ will tend to be close to the theoretical probabilities assigned by the binomial distribution for $n = 20$, $p = \frac{1}{2}$. It is a logical consequence of this and of the normal approximation theorem that the proportion of the X's observed to fall between $20(\frac{1}{2}) + A\sqrt{20 \cdot \frac{1}{2} \cdot \frac{1}{2}} + \frac{1}{2}$ and $20(\frac{1}{2}) + B\sqrt{20 \cdot \frac{1}{2} \cdot \frac{1}{2}} - \frac{1}{2}$ is approximately equal to the area under the standard normal curve between A and B. This deduction can be experimentally tested by tossing a coin $20m$ times, for a large number m (for example, $m = 100$).

9.6 EXERCISES

1. Carry out the experimental verification mentioned above as a cooperative group exercise (for example, 100 tosses for each of 20 persons); and for various values of A and B.

2. Here is a way of testing the normal approximation for $p = \frac{1}{3}$. Take a balanced die: consider the scores 1 and 2 as H and the others as T; the successive tosses of the die generate a sequence of "coin tosses" with $p = \frac{1}{3}$. Repeat Exercise 1 for this case.

chapter 10

CLASSICAL PROBABILITY DISTRIBUTIONS
AND THEIR APPLICATIONS

A few of the classical probability distributions—the geometric, Poisson, and normal—are described, and some of their applications are presented. Each is shown to originate as an approximation to the distribution of a random variable for a game of a large number of coin tosses. The proof of the Poisson approximation is omitted; and the normal approximation depends on the results of Chapter 4, which were given without proof. These approximations are used only in the numerical illustrations and exercises and not in subsequent theoretical material.

10.1 THE GEOMETRIC DISTRIBUTION

Consider the random trial of n tosses of a coin. Let X be the random variable representing the *number of tosses preceding the first* H; we make the convention that X assigns the number 0 to the outcomes in the event "the first H appears on the first toss" and the number n to the outcomes in the event "H does not appear on any of the n tosses." Here is an illustration of the case $n = 4$:

outcome	assigned number	outcome	assigned number
[H H H H]	0	[T H H H]	1
[H H H T]	0	[T H H T]	1
[H H T H]	0	[T H T H]	1
[H T H H]	0	[T H T T]	1
[H T T T]	0	[T T H H]	2
[H T T H]	0	[T T H T]	2
[H T H T]	0	[T T T H]	3
[H H T T]	0	[T T T T]	4

In this particular case, the probability distribution of X is:

$$\Pr(X = 0) = p, \quad \Pr(X = 1) = pq, \quad \Pr(X = 2) = pq^2,$$
$$\Pr(X = 3) = pq^3, \quad \Pr(X = 4) = q^4.$$

In the general case of n tosses, the probability distribution is given as follows:

Proposition 10.1 The probability distribution of X is:

$$\Pr(X = 0) = p, \quad \Pr(X = k) = pq^k, \quad k = 1, \ldots, n - 1,$$
$$\Pr(X = n) = q^n. \tag{10.1}$$

PROOF

1. The event $\{X = 0\}$ has probability p.

Reason. It is the same as the event "H on the first toss."

2. The event $\{X = n\}$ has probability q^n.

Reason. It is the same as the event "T on all n tosses."

3. The event $\{X = k\}$, for $0 < k < n$, is determined by the first $k + 1$ tosses.

Reason. It consists of all outcomes for which T appears on the first k tosses and H on the $(k + 1)$st.

4. The probability of the event "H appears for the first time on the $(k + 1)$st of $k + 1$ tosses" is equal to pq^k.

Reason. The event consists of a single outcome: a multiplet of $k + 1$ letters in which the first k are T and the last H.

5. The event $\{X = k\}$, for k strictly between 0 and n, has probability pq^k.

Reason. Statements 3 and 4, and the Consistency Theorem. ◀

    The proof is complete. We remark that, when n is large, the last probability $\Pr(X = n)$ is very small, so that the probability distribution varies little with n.

Proposition 10.2 *Let X have the probability distribution (10.1); then,*

$$E(X) = q + q^2 + \cdots + q^n. \qquad (10.2)$$

PROOF

1. $E(X) = 0 \cdot p + pq + 2pq^2 + 3pq^3 + \cdots + (n-1)pq^{n-1} + nq^n.$

Reason. Definition 7.4.

2. $E(X) = 0 \cdot p + 1(q - q^2) + 2(q^2 - q^3) + 3(q^3 - q^4)$
$$+ \cdots + (n-1)(q^{n-1} - q^n) + nq^n.$$

Reason. Put $p = 1 - q$ in Statement 1.

3. Formula (10.2) follows from Statement 2.

Reason. Apply the distributive law to the differences within the parentheses; the sum on the right-hand side is

$$q - q^2 + 2q^2 - 2q^3 + 3q^3 - 3q^4 + \cdots + (n-1)q^{n-1} - (n-1)q^n + nq^n,$$

which is equal, upon cancellation of like terms with opposite signs, to the right-hand side of formula (10.2). ◀

Proposition 10.3 $E(X) \quad = \quad (q^{n+1}/p)$
$$q + q^2 + \cdots + q^n = (q/p) - (q^{n+1}/p)$$

PROOF. The right-hand side of this equation is $(q - q^n)/(1 - q)$. Multiply both sides by $1 - q$ and then compare them. ◀

Proposition 10.4 *Consider the sequence of expected values formed by putting $n = 1, 2, \ldots$ in formula (10.2). The limit of this sequence is (q/p); in other words, if n is very large, then we have, approximately,*

$$E(X) = q/p. \qquad (10.3)$$

PROOF

1. The limit of the sequence $q/p, q^2/p, \ldots, q^n/p, \ldots$ is 0.

Reason. Definition 3.3.

2. The limit of the sequence $(q/p) - (q^1/p), (q/p) - (q^2/p), \ldots, (q/p) - (q^n/p), \ldots$ is q/p.

Reason. Statement 1 and Definition 3.5.

3. The assertion of the proposition follows.

Reason. Statement 2 and Propositions 10.1 and 10.2.   ◄

The probability distribution formally obtained by putting $n = \infty$ in formula (10.1) is known as the geometric distribution. Its graph for selected values of p is shown in Fig. 10.1. The expected value of a random variable X having this probability distribution is, also formally, given by Eq. (10.3). Here are some applications.

1) *Length of telephone conversations.* It has been found by experience that the length of a telephone conversation is like a random variable with a geometric distribution; more exactly, if a conversation is selected at random from an ensemble of conversations, and if the unit of length (in time) is prescribed, then the probability distribution of the length of the conversation is that of the geometric distribution for some prescribed value p. This experimental fact can be explained in terms of the theory of the coin game. During the course of a conversation the participants are usually unaware of how long they have spoken: the decision to terminate the conversation at any moment is independent of the length of the conversation up to that moment. We can think to the *time of termination* as being determined by successive tosses of a coin: the conversation is maintained as long as T's appear, and is terminated at the first appearance of H; thus, the length of the conversation (or equivalently, the time of termination) is like the number of tosses preceding the first H.

2) *Lifetimes of mechanical parts.* It has been found that the lifetimes of mechanical components are like random variables with geometric distributions; more specifically, this is true of components which do not "age" with use. Electrical equipment, such as fuses, seem to have this property. That the lifetimes have the geometric distribution can be explained in terms of the coin game. A fuse fails at a time when the circuit is just turned on. Each activation of the circuit can be thought of as the toss of a coin: the outcomes

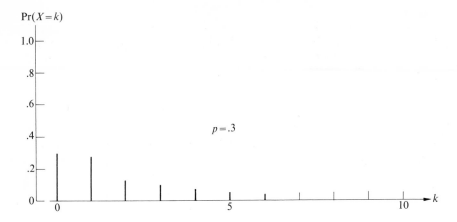

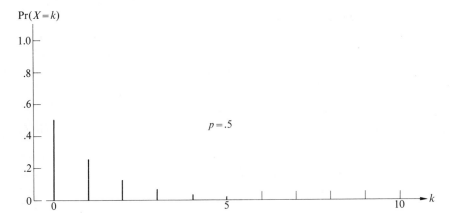

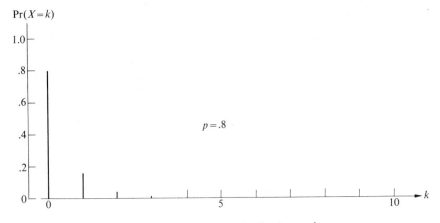

Fig. 10.1. Geometric distribution: pq^k.

H and T correspond to failure and nonfailure of the fuse, respectively. The fuse lasts as long as T turns up on each toss; it fails as soon as H appears; hence, the life of the fuse has the same distribution as the number of tosses up to the first H, the geometric distribution.

10.2 THE POISSON DISTRIBUTION

We shall now consider an approximation to the binomial distribution when n is very large, p very small, and the mean np of moderate size. Under these conditions, the probabilities in the binomial distribution can be approximated by the corresponding probabilities of the probability distribution known as the Poisson distribution, named after S. D. Poisson (1781–1840).* Though the binomial distribution depends on two parameters, n and p, the Poisson distribution contains the single parameter λ. In approximating the former by means of the latter distribution, we take λ to be equal to np, which is the expected Number of H's in n tosses; λ represents the expected value of a random variable having the Poisson distribution. There is a simple algebraic formula for the general term of the Poisson distribution: if X has the Poisson distribution, then

$$\Pr(X = k) = e^{-\lambda}\lambda^k/k!, \qquad k = 0, 1, \ldots; \qquad (10.4)$$

however, this formula is now of theoretical rather than practical interest because the values of (10.4) have been thoroughly tabulated. Table III contains the values of the Poisson distribution for parameter values of λ from 1 to 10, inclusive. The graph of the Poisson distribution is shown for various values of λ in Fig. 10.2.

Here are several illustrations of the use of the Poisson distribution.

3) *Fire insurance.* A home fire insurance company has a large number n of policy holders. The probability that any particular home is destroyed as a result of fire during a given year is a very small number p. The destruction of one home by fire does not affect (in most cases) that of any other home. On the strength of these assumptions we can think of the net income of the insurance company as the fortune of a gambler tossing a coin with a small probability p a large number of times n. Each toss corresponds to a policy holder: the outcome H corresponds to his home burning down during the given year, and the outcome T to the home surviving. To simplify the model, we shall consider only fires which result in almost total loss of the

---

* See W. Feller, *An Introduction to Probability Theory and its Applications*, 3rd ed., Vol. 1, New York: Wiley, 1968, p. 153.

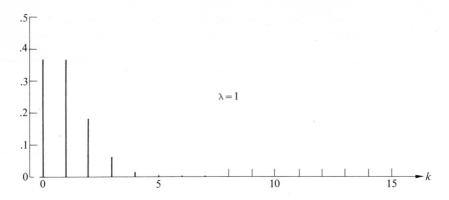

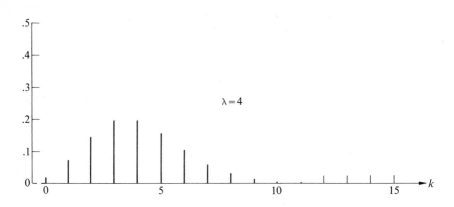

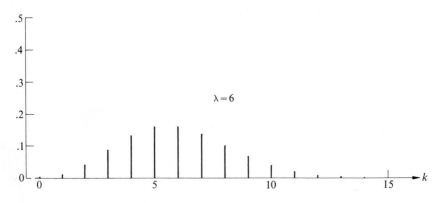

Fig. 10.2. Poisson distribution.

home, not just in partial damage. The number of homes destroyed during the year is the Number of H's in n tosses of a coin, where n is large and p small. The probability distribution can be approximated by the Poisson distribution. The insurance company can use these probabilities to compute the minimum insurance rates that will provide it with a high probability of covering policy claims.

Suppose there are $n = 10{,}000$ policies and that during a one-year period each home has probability .0005 of burning; here

$$\lambda = np = (10{,}000)(.0005) = 5.$$

The probabilities that exactly 0, 1, 2, ..., 15 homes burn are given in the respective entries of column 5 in Table III. The table has been calculated so that the sum of the probabilities corresponding to numbers greater than 15 has a sum less than .0001; even though such numbers bear positive probabilities, these are negligible. Suppose that each home is insured for $20,000. If the company sought to cover *all* possible fire losses with absolute certainty, then it would have to charge each policy holder a premium equal to the value of his home; however, no one would buy such insurance. So the company must reduce the premium and bear a small *probability* of not being able to pay for all fire losses. The probability that more than 12 homes are destroyed during the year is the sum of the probabilities for 13, 14, 15, 16, ... ; according to the last three entries in column 5 of Table III the probability is $.0013 + .0005 + .0002 = .0020$. The loss of 12 homes results in a total liability of $20{,}000 \cdot 12 = \$240{,}000$; hence, if each policy holder pays an equal share of this quantity, or $24, the company will be able to cover all fire losses with probability .998. There is a large probability that the company will have a surplus at the end of the year. Let x be the number of lost homes; the company earns $20{,}000 \cdot (12 - x)$; therefore, there is probability .0082 that $20,000 is earned, probability .0181 that $40,000 is earned, etc.

4) *Cases of a rare disease.* Certain diseases occur in a very small proportion p of a population. These are usually not communicable, so that the presence of the disease in one person does not affect its presence in any other person. We can imagine the toss of a coin corresponding to each person: H and T are associated with the presence and absence of the disease for that person. The number of cases in a large group of n persons is like the Number of H's in n tosses of a coin with a small probability p of H on each toss; its probability distribution is approximately given by the Poisson distribution with $\lambda = np$. Suppose that a disease is known to occur in one person per thousand in a population; then the probabilities of exactly 0, 1, ...

cases in a town of 3000 people are given by the respective entries of Table III in the column $\lambda = 3$. The probability of at least two cases, which is 1 minus the probabilities of no cases and exactly one case, respectively, is equal to $1 - .0498 - .1494 = .8008$.

5) *Industrial acceptance sampling.* We refer to the application in Section 2.2D. There it was shown that, if a large lot of manufactured items has a proportion p of defectives, and if n items are selected for inspection, then the number of defectives among the latter is like the Number of H's in n tosses of a coin. In many cases of practical interest, the proportion p is very small, and the number of items n inspected so large that np is of moderate size; thus, the probability distribution of the number of defective items is approximately equal to the Poisson distribution with parameter $\lambda = np$. This approximation may be used in the construction of the operating characteristic $L(p)$ for small p.

Here is a numerical illustration. Let us find the operating characteristic of the sampling plan under which 100 items are selected for inspection, and the lot is rejected if more than two items are found to be defective; for this plan we have $n = 100$ and $c = 2$. Let us determine $L(p)$ for $p = .01, .02, .03, .04,$ and $.05$. Expression (2.6) with $p = .01$, $n = 100$, and $c = 2$ is approximated by the sum of the first three terms of the Poisson distribution with $\lambda = (.01)(100) = 1$; it is $.3679 + .3679 + .1839 = .9197$. For $p = .02$, we find the sum of the first three terms of the Poisson distribution with $\lambda = (.02)(100) = 2$ to be $.1353 + .2707 + .2707 = .6767$; similarly, for $p = .03, .04,$ and $.05$, we find the sums to be $.4232, .2381,$ and $.1246$, respectively. In summary, the values of $L(p)$ for the five corresponding values of p are $.9197, .6767, .4232, .2381,$ and $.1246$. These probabilities indicate that the probability of accepting the lot decreases rapidly as the proportion of defectives increases from $p = .01$ to $p = .05$.

10.3 THE NORMAL DISTRIBUTION

The standard normal curve was introduced in Chapter 4 as a tool in the approximation of sums of terms of the binomial distribution. In this section we shall show why this curve has found use in so many different areas of application.

Let us recall the random walk in Chapter 6, supposing, for simplicity, that it is *symmetric*, that is, that $p = \frac{1}{2}$; here, Cain and Abel play with a balanced coin (cf. Example 6.5). Let S_n be the random variable representing Abel's fortune after n tosses; if n is even, then the set of values of S_n is the set of *even* integers $0, \pm 2, \pm 4, \ldots, \pm n$; and if n is odd the set of values of S_n is the set of *odd* integers $\pm 1, \pm 3, \ldots, \pm n$. The normal approximation to the binomial distribution provides an approximation to the probability distribution of S_n when n is large.

> **Proposition 10.5** *The probability that S_n, Abel's fortune after n tosses, is between numbers*
>
> $$a = \sqrt{n}\, A + 1 \quad and \quad b = \sqrt{n}\, B - 1, \qquad (10.5)$$
>
> *inclusive, is approximately equal to the area under the standard normal curve between A and B if n is large; more precisely, if P_n is the above probability for n tosses, then the limit of the sequence $P_1, P_2, \ldots, P_n, \ldots$ is the indicated area under the curve.*

PROOF. The proof of this proposition is a direct consequence of the normal approximation to the binomial distribution and the following statement:

1. Abel's Fortune is between $a = \sqrt{n}\, A + 1$ and $b = \sqrt{n}\, B - 1$ if and only if the Number of H's is between

$$(n/2) + \sqrt{n/4}\, A + \tfrac{1}{2} \quad and \quad (n/2) + \sqrt{n/4}\, B - \tfrac{1}{2}.$$

Reason. Abel's Fortune S_n is the *difference* between the Number of H's and the Number of T's; and the Number of H's *plus* the Number of T's is n; therefore, the Fortune is equal to 2(Number of H's) $- n$; and, finally, the latter is between $\sqrt{n}\, A + 1$ and $\sqrt{n}\, B - 1$ if and only if the Number of H's is between $\tfrac{1}{2}(n + \sqrt{n}\, A + 1)$ and $\tfrac{1}{2}(n + \sqrt{n}\, B - 1)$. ◀

Example 10.1 After $n = 100$ tosses, the probability that Abel's Fortune is between -19 and $+9$ is obtained by solving Eq. (10.5) for A and B and finding the appropriate area under the standard normal curve. From $-19 = \sqrt{100}\, A + 1$ and $9 = \sqrt{100}\, B - 1$ we get $A = -2$ and $B = +1$. The area between -2 and $+1$ is, in accordance with the rules of Chapter 4, equal to $.4772 + .3413 = .8185$. ◁

Proposition 10.5 motivates the definition of the *normalized* Fortune $S_n/\sqrt{n}$: this is the random variable representing Abel's Fortune divided by the square root of the number of tosses; more specifically, if an outcome of the game of n tosses is represented by a net fortune of x dollars for Abel, then that outcome is represented by a normalized net fortune of $x/\sqrt{n}$ dollars. The following is a corollary of Proposition 10.5.

Proposition 10.6 *The probability that* $S_n/\sqrt{n}$, *Abel's normalized Fortune after n tosses, is between* $A + (1/\sqrt{n})$ *and* $B - (1/\sqrt{n})$ *is approximately equal to the area under the standard normal curve between A and B, if n is large.*

PROOF. The *normalized Fortune* is between $A + (1/\sqrt{n})$ and $B - (1/\sqrt{n})$ if and only if the *Fortune* is between $a = \sqrt{n}\,A + 1$ and $b = \sqrt{n}\,B - 1$; hence, this proposition follows from the previous one.   ◀

As a consequence of Proposition 10.6 we shall say that the normalized Fortune $S_n/\sqrt{n}$ has an *approximate standard normal distribution* when n is large.

Now we shall define a *normal distribution*, which is more general than the *standard* normal distribution considered up to now. The set of values of the random variable $S_n/\sqrt{n}$ is determined by the money won or lost on each play (the monetary scale); for example, if two dollars instead of one dollar are transferred after each toss, then the fortune in the former case is twice as large as in the latter case. More generally, if σ is the number of monetary units transferred on each toss, then the fortune is σ times what it would have been if one unit had been transferred on each toss; in this case, the probability that the normalized fortune $S_n/\sqrt{n}$ is between $A + (\sigma/\sqrt{n})$ and $B - (\sigma/\sqrt{n})$ is the area under the standard normal curve between A/σ and B/σ.

Example 10.2 Consider the game in which Cain and Abel win or lose two dollars on each toss. Let us find the probability that Abel's Fortune is between -20 and $+10$ (inclusive) after 100 tosses or, equivalently, that the normalized Fortune is between -2 and $+1$; here

$$-2 = A + 2/\sqrt{100},$$
$$+1 = B - 2/\sqrt{100},$$

or $A = -2.2$ and $B = +1.2$; thus, A/σ is equal to -1.1 and B/σ to .6. The area under the standard normal curve between -1.1 and .6 is

$$.3643 + .2257 = .5900,$$

which is the desired probability.   ◁

Definition 10.1 If σ monetary units are won or lost on each toss, then the normalized Fortune $S_n/\sqrt{n}$ is said to have an *approximate normal distribution with standard deviation* σ.

The *normal curve* with standard deviation σ is the graph of the function

$$y = (1/\sqrt{2\pi}\,\sigma)e^{-(x^2/2\sigma^2)};$$

the latter is obtained from the function in formula (4.1) by replacing x and y by x/σ and $y\sigma$, respectively. The influence of σ on the form of the graph is illustrated in Fig. 10.3.

Suppose that Abel starts the game with an initial fortune of m dollars, where m is not necessarily 0 but may be positive or negative; assume that σ is the number of monetary units won or lost on each play. Abel's Fortune after n plays is S_n (net Fortune won) plus m, the initial fortune; thus the normalized Fortune is $(S_n + m)/\sqrt{n}$. Now define the number μ as

$$\mu = m/\sqrt{n};$$

then, $m = \mu\sqrt{n}$. The probability that the normalized Fotune $(S_n + m)/\sqrt{n}$ is between $A + \sigma/\sqrt{n}$ and $B - \sigma/\sqrt{n}$ is equal to the probability that the *net* Fortune, divided by $\sqrt{n}$, $S_n/\sqrt{n}$, is between

$$A - \mu + \frac{\sigma}{\sqrt{n}} \quad \text{and} \quad B - \mu - \frac{\sigma}{\sqrt{n}}.$$

This is given approximately by the area under the standard normal curve between $(A - \mu)/\sigma$ and $(B - \mu)/\sigma$.

Definition 10.2 If σ monetary units are won or lost on each toss, and if Abel starts with an initial fortune of $m = \mu\sqrt{n}$ units, then Abel's fortune after n tosses, divided by the square root of n, has the approximate *normal distribution with mean μ and standard deviation σ.*

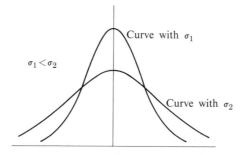

Figure 10.3

The *normal curve* with mean μ and standard deviation σ is obtained from the normal curve with standard deviation σ (defined above) by displacing it μ units to the right if μ is positive or $-\mu$ units to the left if μ is negative. This curve is also bell-shaped and it is symmetric with respect to the point μ. When $\mu = 0$ and $\sigma = 1$, the normal curve reduces to the standard one.

Example 10.3 Consider the game in which Cain and Abel win or lose five dollars on each toss and where Abel starts with an initial fortune of $m = 40$ units. Suppose there are 100 tosses; then $\mu = \frac{40}{10} = 4$ and $\sigma = 5$. The probability that Abel's normalized Fortune is between $A + \sigma/\sqrt{n}$ and $B - \sigma/\sqrt{n}$ is given approximately by the area under the standard normal curve between

$$(A - 4)/5 \quad \text{and} \quad (B - 4)/5;$$

for example, the probability that Abel's normalized Fortune is between -1.5 and $+4.5$ is the area between

$$(-2 - 4)/5 = -1.2 \quad \text{and} \quad (5 - 4)/5 = .2,$$

which is $.3849 + .0793 = .4642.$ ◁

Here are some applications of the normal distribution.

6) *Theory of errors.* It has been observed that, when a physical quantity is repeatedly measured by a measuring instrument, the successive numbers produced are different one from another; furthermore, when arranged in a frequency distribution, they usually seem to produce a symmetric bell-shaped frequency curve. This is the experimental foundation of the *law of errors*. There is a *mathematical* justification for this law based on the coin-tossing game and the normal approximation. The measurement of a physical quantity such as the weight of a chemical substance or the intensity of a light source can be considered as the total normalized fortune of a player in a coin-tossing game. The actual number yielded by a measurement is equal to the true constant under measurement plus the sum of many positive and negative small and independent deviations—"random errors"—due to the operation and condition of the measuring instrument. These small random errors may be considered to be the successive amounts won (or lost) by Abel in a coin game with a balanced coin; the sum of these errors is like the net Fortune won by Abel, divided by the square root of n; and the *true* constant under measurement is like the initial fortune divided by $\sqrt{n}$. (The division by $\sqrt{n}$ is introduced for the purpose of putting the fortune on a standard scale.) It follows that the number produced by a measurement has a distribution given (approximately) by a normal distribution whose mean μ is the true constant being measured and whose standard deviation σ is determined by the units of measurement and by the accuracy of the device used.

The Clustering Principle of Chapter 9 now explains the bell-shaped frequency distribution of successive measurements. The measurements are independent random variables with a common probability distribution which is approximately normal; by virtue of the above Principle, the empirical frequencies of the measurements tend to cluster about the theoretical probabilities of their common (normal) distribution.

7) *Biological characteristics.* When some biological characteristic of each person of a large group is measured, and the resulting numbers are arranged in a frequency distribution, it is often found that the frequencies have the familiar bell-shaped symmetric pattern of the normal curve; this has been found to be true for heights, weights, intelligence quotients, and many other characteristics. This has a theoretical explanation in terms of the coin game. The observed characteristic of a person can be imagined to be the sum of a universal constant common to all people plus the sum of many small positive and negative independent "random errors"; the latter sum is like the normalized net Fortune won by a player in a coin-tossing game. It follows that the probability distribution of the observed measurement of a personal characteristic has an approximate normal distribution with a mean equal to the universal constant and some variance determined by the scale of measurement. The Clustering Principle explains the appearance of the normal curve in the empirical frequency distribution.

EXERCISES FOR CHAPTER 10

1. A coin with $p = .4$ is tossed 10 times. Find the probability that the first H appears (a) on the fourth toss, (b) after the eighth toss, (c) before the third toss.

2. A coin with $p = .6$ is tossed four times. Let X be the number of tosses preceding the first H. Find the variance of X.

3. Carry out the proof of Proposition 10.2 in the particular case $n = 5$.

4. Perform the multiplication in the proof of Proposition 10.3 in the particular case $n = 6$.

5. Using Propositions 10.2, 10.3, and 10.4, estimate the difference between $E(X)$ and q/p in the case $p = .5$, $n = 10$; in the case $p = .4$, $n = 12$.

6. It can be shown that the Poisson distribution can be approximated by the normal distribution when λ is very large; find the justification for this.

7. A fire insurance company has 20,000 policies, insuring each holder's home for $20,000. The probability of the destruction of a home by fire is .0003. What premium must each holder be charged in order that the probability that all losses be covered be at least .9995?

8. Repeat Exercise 7 for 50,000 policies.

9. A disease occurs in one person per 100,000 in the population. What is the probability of finding more than eight cases in a city of 300,000? A city of 500,000?

10. A disease occurs in one person per 10,000. What is the probability of at least 10 cases in a population of 40,000? At least three cases?

11. Fifty items are selected for inspection from a large lot. The lot is rejected if four or more are found to be defective. Find the operating characteristic $L(p)$ for $p = .02, .04, .06, .1$.

12. Repeat Exercise 11 for a sample of 100 items.

13. A balanced coin is tossed 625 times. Find the probability that Abel's Fortune is greater than 20; and the probability that his normalized Fortune is between $-.3$ and $+.8$.

14. Cain and Abel play with a balanced coin; six dollars are won or lost on each toss. If Abel starts with 15 dollars, what is the probability that his normalized Fortune after 25 tosses is between 1.5 and 3.5?

15. Suppose, in the previous exercise, that two dollars are won or lost on each toss, and that Abel starts with $16. Find the probability that his normalized Fortune after 64 tosses is between 3 and 5.

16. Perform the following computational experiment. Take the first 200 stocks listed on the New York Stock Exchange, and record the net gains for the past day; these may be positive, negative, or zero. Arrange the net gains in a frequency distribution; find the numbers of stocks that gained 0, $\pm(\frac{1}{8})$, $\pm(\frac{2}{8})$, . . . , respectively. What is the shape of the distribution obtained?

17. Toss a coin until H appears, recording the number of preceding T's. Perform this 25 times, arranging the results in a frequency distribution. Compare it to the geometric distribution.

chapter 11

SUMS AND PRODUCTS
OF RANDOM VARIABLES

The best-known results of classical probability theory are about sums of independent random variables. Now we define the notions of sums and products of random variables, and a particularly useful random variable—the indicator. In Section 11.4 an important formula for the sum of a random number of random variables is derived, and is applied in Section 11.5 to the "branching process."

11.1 SUMS OF RANDOM VARIABLES AS RANDOM VARIABLES

Definition 11.1 Let $[X, Y]$ be a composite random variable on a random trial (see Chapter 8 and Definition 7.6). The *sum* of X and Y, denoted $Z = X + Y$, is defined as the random variable which associates with each outcome of the random trial the sum of the values assigned by X and Y, respectively: if x and y are the values associated with a particular outcome by X and Y, respectively, then $x + y$ is the value associated with it by the random variable Z.

Example 11.1 A coin is tossed four times. Let X and Y be the Numbers of H's appearing on the first two and last two tosses, respectively. Put $Z = X + Y$; then Z is the Number of H's appearing on all four tosses; for example, if one H appears on the first two tosses ($X = 1$) and two H's on the last two ($Y = 2$), then $Z = 1 + 2 = 3$.   ◁

Example 11.2 Let X and Y be the Scores on two tosses of a die; then $Z = X + Y$ is the sum of the Scores. It is a random variable with the probability distribution displayed in formula (8.4).   ◁

> **Proposition 11.1** *The probability distribution of $Z = X + Y$ can be calculated from the joint probability distribution of $[X, Y]$: for any value z of Z, the probability $\Pr(Z = z)$ is equal to the sum of all probabilities $\Pr(X = x, Y = y)$ for all values $[x, y]$ of $[X, Y]$ satisfying the equation $z = x + y$.*

PROOF

1. Let $x_1, x_2, \ldots$ and $y_1, y_2, \ldots$ be the values of X and Y, respectively, which satisfy the equations $z = x_1 + y_1$, $z = x_2 + y_2$, $\ldots$; then the event $\{Z = z\}$ is the union of the disjoint events $\{X = x_1, Y = y_1\}$, $\{X = x_2, Y = y_2\}, \ldots$.

Reason. (See Definition 7.7.) By virtue of Definition 11.1, an outcome is represented by the value z of the random variable Z if and only if it is represented by one of the pairs of values $[x_1, y_1], [x_2, y_2], \ldots$ of the composite random variable $[X, Y]$.

2. $\Pr(Z = z) = \Pr(X = x_1, Y = y_1) + \Pr(X = x_2, Y = y_2) + \cdots$.

Reason. This follows from Statement 1, Proposition 5.2, and the definition of $\Pr(Z = z)$.   ◀

Example 11.3 Let X and Y be the Scores on two tosses of a die, and Z the sum of the Scores; then (cf. Section 8.2)

$$\Pr(Z = 2) = \Pr(X = 1, Y = 1) = \tfrac{1}{36},$$
$$\Pr(Z = 3) = \Pr(X = 1, Y = 2) + \Pr(X = 2, Y = 1) = \tfrac{2}{36},$$
$$\Pr(Z = 4) = \Pr(X = 1, Y = 3) + \Pr(X = 2, Y = 2)$$
$$+ \Pr(X = 3, Y = 1) = \tfrac{3}{36}, \text{ etc.} \qquad ◁$$

Example 11.4 Let X_1 and X_2 be the random variables considered in Example 7.4; the joint probability distribution is given following Definition 7.8.

The probability distribution of the random variable $Z = X_1 + X_2$ is calculated as follows:

The event $\{Z = 0\}$ is the event $\{X_1 = 0, X_2 = 0\}$; hence,

$$Pr(Z = 0) = q^3.$$

The event $\{Z = 2\}$ is the event $\{X_1 = 1, X_2 = 1\}$; hence,

$$Pr(Z = 2) = 3pq^2.$$

And so on.   ◁

The following proposition relates the expected value of a sum of random variables to the expected values of the latter.

Proposition 11.2 *Let* $[X, Y]$ *be the composite random variable on a random trial, and* $Z = X + Y$; *then* $E(Z) = E(X) + E(Y)$.

PROOF

1. $E(Z)$ is the sum of all terms $z \cdot Pr(Z = z)$.

Reason. Definition 7.4.

2. The sum in Statement 1 is equal to the sum of all terms

$$(x + y)Pr(X = x, Y = y).$$

Reason. Every value z of Z is of the form $x + y$, by Definition 11.1; and every probability $Pr(Z = z)$ is the sum of probabilities $Pr(X = x, Y = y)$, where $z = x + y$, by Proposition 11.1.

3. The sum in Statement 2 may be split into two distinct sums: the sum of all terms $x \cdot Pr(X = x, Y = y)$, and the sum of all terms $y \cdot Pr(X = x, Y = y)$.

Reason. Apply the distributive law to each summand:

$$(x + y)Pr(X = x, Y = y) = x \cdot Pr(X = x, Y = y) + y \cdot Pr(X = x, Y = y).$$

4. For each value x, the sum of the probabilities $Pr(X = x, Y = y)$ over all the y-values is equal to $Pr(X = x)$.

Reason. We apply Proposition 5.5. Let $y_1, y_2, \ldots$ be the set of values of Y, and $E_1, E_2, \ldots$ the events $\{Y = y_1\}$, $\{Y = y_2\}$, $\ldots$, respectively. These are disjoint, and their union is the sure event; therefore, they satisfy the hypothesis of Proposition 5.5. Let A be the event $\{X = x\}$, and $A_1, A_2, \ldots$ the events $\{X = x, Y = y_1\}$, $\{X = x, Y = y_2\}$, $\ldots$, respectively; the latter are the intersections of A with $E_1, E_2, \ldots$, respectively, by Definition 7.7. The conclusion of Proposition 5.5 implies that $\Pr(X = x)$ is equal to the sum $\Pr(X = x, Y = y_1) + \Pr(X = x, Y = y_2) + \cdots$.

5. The sum of the terms $x \cdot \Pr(X = x, Y = y)$ over all values x and y is equal to $E(X)$.

Reason. The sum may be calculated by summing over y for each value of x, and then summing over all values of x. Let $[x_1, y_1], [x_1, y_2], \ldots ; [x_2, y_1], [x_2, y_2], \ldots ; \ldots$ be the values of $[X, Y]$. The sum in the Statement may be written as

$$x_1 \cdot \Pr(X = x_1, Y = y_1) + x_1 \cdot \Pr(X = x_1, Y = y_2) + \cdots$$
$$+ x_2 \cdot \Pr(X = x_2, Y = y_1) + x_2 \cdot \Pr(X = x_2, Y = y_2) + \cdots$$
$$+ \cdots .$$

By virtue of Statement 4, the sums of the terms in the *rows* are

$$x_1 \cdot \Pr(X = x_1), \qquad x_2 \cdot \Pr(X = x_2), \qquad \ldots ,$$

respectively; therefore, the sum of the row sums is $E(X)$ in accordance with Definition 7.4.

6. The sum of the terms $y \cdot \Pr(X = x, Y = y)$ over all values of x and y is equal to $E(Y)$.

Reason. Interchange the roles of X and Y and of x and y and apply the conclusion of Statement 5.

7. The conclusion of the Proposition is valid: $E(Z) = E(X) + E(Y)$.

Reason. Statements 1, 3, 5, and 6.   ◀

Example 11.5 Let X, Y, and Z be the random variables in Example 11.1. We have, by Proposition 3.1, $E(X) = 2p$, $E(Y) = 2p$, and $E(Z) = 4p$. This illustrates the conclusion of Proposition 11.2.   ◁

Example 11.6 Let X, Y, and Z be the random variables in Example 11.2; then $E(X) = E(Y) = 3.5$. (Verify this.) It was shown at the end of Chapter 8 that the expected value of the sum of two Scores, $E(Z)$, is 7; thus $E(Z) = E(X) + E(Y)$.   ◁

Example 11.7 Let X_1, X_2, and Z be the random variables considered in Example 11.4, which refers to Example 7.4. The probability distributions are, as given following Definition 7.3,

$$\Pr(X_1 = 0) = q^3, \qquad \Pr(X_1 = 1) = 3q^2p, \qquad \Pr(X_1 = 2) = 3qp^2,$$
$$\Pr(X_1 = 3) = p^3;$$

$$\Pr(X_2 = 0) = q^3, \qquad \Pr(X_2 = 1) = 3q^2p + p^2q, \qquad \Pr(X_2 = 2) = 2p^2q,$$
$$\Pr(X_2 = 3) = p^3.$$

The expected values, as computed after Definition 7.4, are $E(X_1) = 3p$ and $E(X_2) = 3p - p^2 + p^3$. The probability distribution of $Z = X_1 + X_2$ is

$$\Pr(Z = 0) = q^3, \qquad \Pr(Z = 2) = 3pq^2, \qquad \Pr(Z = 3) = p^2q,$$
$$\Pr(Z = 4) = 2p^2q, \qquad \Pr(Z = 6) = p^3.$$

It follows that $E(Z)$ is

$$0 \cdot q^3 + 2 \cdot 3 \cdot pq^2 + 3p^2q + 4 \cdot 2p^2q + 6p^3 = 6p - p^2 + p^3,$$

which is, in fact, equal to $E(X_1) + E(X_2)$.   ◁

All the results given for a composite random variable $[X, Y]$ of two components can be generalized to composite random variables $[X_1, \ldots, X_k]$ of k components.

Definition 11.2 Let $[X_1, \ldots, X_k]$ be a composite random variable on a random trial. The random variable $Z = X_1 + \cdots + X_k$ associates with each outcome of the random trial the sum of the values associated by $X_1, \ldots, X_k$, respectively: if $[x_1, \ldots, x_k]$ is a multiplet associated with an outcome by $[X_1, \ldots, X_k]$, then $x_1 + \cdots + x_k$ is the value associated with it by the random variable Z.

The proofs of the following two propositions contain the same ideas as those of Propositions 11.1 and 11.2, respectively; hence, they are omitted.

Proposition 11.3 *The probability distribution of $Z = X_1 + \cdots + X_k$ can be calculated from the joint probability distribution of $[X_1, \ldots, X_k]$: for any value z of Z, the probability $\Pr(Z = z)$ is equal to the sum of all probabilities $\Pr(X_1 = x_1, \ldots, X_k = x_k)$ for all multiplets $[x_1, \ldots, x_k]$ whose components satisfy the equation $z = x_1 + \cdots + x_k$.*

Proposition 11.4 *Let $[X_1, \ldots, X_k]$ be a composite random variable, and let Z be the random variable $Z = X_1 + \cdots + X_k$; then, $E(Z) = E(X_1) + \cdots + E(X_k)$.*

11.1 EXERCISES

1. Let X and Y be independent random variables with the common probability distribution

value:	0	1	2
probability:	.3	.4	.3

Find the probability distribution of $Z = X + Y$, and then find $E(Z)$ by two methods: first, directly from the probability distribution of Z, and second, by applying Proposition 11.2.

2. Let X_1, X_2, and X_3 be independent random variables with the common probability distribution given in Exercise 1. Put $Z = X_1 + X_2 + X_3$ and proceed as in Exercise 1.

3. Let $[X, Y]$ be a composite random variable with the joint probability distribution

value	probability	value	probability	value	probability
[0, 0]	.2	[1, 0]	.1	[2, 0]	.05
[0, 1]	.1	[1, 1]	.1	[2, 1]	.1
[0, 2]	.1	[1, 2]	.05	[2, 2]	.2

a) Find the probability distributions of X and Y, respectively; and $E(X)$ and $E(Y)$.

b) Find the probability distribution of $Z = X + Y$, and then compute $E(Z)$. Compare the latter with the sum $E(X) + E(Y)$ obtained in (a).

4. A gambler tosses a balanced coin three times. He wins one dollar if H appears on the first toss, and nothing if T appears; he wins two dollars for H on the second toss, and nothing for T; and he wins three dollars for H on the third toss, and nothing for T. Let X_1, X_2, and X_3 represent his winnings on the three respective tosses, and put $Z = X_1 + X_2 + X_3$. Find the probability distribution of Z, and compute $E(Z)$.

5. A gambler tosses a die twice. He is paid an amount equal to the *square* of the Score on each toss. Let Z be the sum of the squares of the two Scores; find $E(Z)$ using Proposition 11.2.

6. Using Proposition 11.1 and formula (8.4), find the probability that the sum of the Scores on four tosses of a die is 14.

7. Show how the expression np for the expected number of H's in n tosses of a coin (Proposition 3.1) can be derived as a consequence of Proposition 11.4. (Hint: Express the Number of H's as a sum of n random variables.)

8. Show by mathematical induction on k that Proposition 11.4 is a *consequence* of Proposition 11.2: show that the relation

$$E(X_1 + \cdots + X_{k-1}) = E(X_1) + \cdots + E(X_{k-1})$$

for *every* set of $k - 1$ random variables and Proposition 11.2 together imply that $E(X_1 + \cdots + X_k) = E(X_1) + \cdots + E(X_k)$ for every set of k.

11.2 INDICATOR RANDOM VARIABLES

Definition 11.3 A random variable X with a set of values containing just the two numbers 0 and 1 is called an indicator random variable.

Suppose that X is an indicator random variable, and let A be the event $\{X = 1\}$; then the complement of A is the event $\{X = 0\}$, and the equation $\Pr(X = 1) = 1 - \Pr(X = 0)$ follows from Proposition 5.3.

Proposition 11.5 *If X is an indicator random variable, then* $E(X) = \Pr(X = 1)$.

PROOF. By Definition 7.4, we have

$$E(X) = 0 \cdot \Pr(X = 0) + 1 \cdot \Pr(X = 1) = \Pr(X = 1). \qquad \blacktriangleleft$$

Proposition 11.6 *Let $X_1, \ldots, X_k$ be indicator random variables on a random trial; put $Z = X_1 + \cdots + X_k$; then*

$$E(Z) = \Pr(X_1 = 1) + \cdots + \Pr(X_k = 1).$$

PROOF. This is an immediate consequence of Propositions 11.4 and 11.5. ◀

Here is an application of indicator random variables.

Number of returns of the Fortune to 0 in coin game. Consider the coin-tossing game between Cain and Abel in which one dollar is transferred after each toss. The Fortune is 0 at the kth toss if the Number of H's is equal to the Number of T's in the first k tosses; in such a case, the integer k is necessarily an *even* integer.

Let us consider the particular case of a game of six tosses. It is possible for the Fortune to be 0 on the second, fourth, and sixth tosses. Let Z be the random variable representing the number of returns to 0; then Z has the set of values 0, 1, and 2. Though it is laborious to compute the probability distribution of Z, it is relatively easy to find $E(Z)$ by the use of indicator random variables and the application of Proposition 11.6. Let X_1, X_2, and X_3 be the following random variables: The variable X_1 assigns the value 1 to all outcomes of the six tosses for which the Fortune is equal to 0 on the second toss, and assigns the value 0 to all other outcomes—in other words, the event $\{X_1 = 1\}$ is the event "the Fortune is 0 on the second toss," and $\{X_1 = 0\}$ is its complement. The variables X_2 and X_3 are analogously defined by the relations: $\{X_2 = 1\}$ is the event "the Fortune is 0 on the fourth toss," and $\{X_2 = 0\}$ is its complement; and $\{X_3 = 1\}$ is the event "the Fortune is 0 on the sixth toss," and $\{X_3 = 0\}$ is its complement. These three random variables are indicators. Their sum $X_1 + X_2 + X_3$ is the random variable Z representing the number of times the Fortune returns to 0. The expected value of Z can, by virtue of Proposition 11.6, be expressed as

$$E(Z) = \Pr(X_1 = 1) + \Pr(X_2 = 1) + \Pr(X_3 = 1).$$

The first term on the right-hand side of this equation is, by definition, the probability that the Number of H's is equal to the Number of T's on the first two tosses, or, in other words, that the Number of H's is 1; this probability is $2pq$. The probabilities $\Pr(X_2 = 1)$ and $\Pr(X_3 = 1)$ are the probabilities of exactly 2 H's in the first four tosses and of exactly 3 H's in the first six tosses; these are

$$\binom{4}{2} p^2 q^2 \quad \text{and} \quad \binom{6}{3} p^3 q^3,$$

respectively. We conclude that $E(Z)$ is given by the formula

$$E(Z) = 2pq + 6p^2 q^2 + 20p^3 q^3.$$

In the particular case of a balanced coin ($p = \frac{1}{2}$) we find that $E(Z) = 2(\frac{1}{4}) + 6(\frac{1}{4})^2 + 20(\frac{1}{4})^3 = \frac{19}{16}$. It follows from Proposition 3.3 that $E(Z)$ is *largest* in the special case of a balanced coin. This is consistent with intuitive considerations: if the coin is more likely to turn up H (or T) on each toss, then Abel's (or Cain's) Fortune is more likely to be positive.

The construction above can be carried out for a game with an arbitrary number of tosses.

11.2 EXERCISES

1. Find the expected number of returns of the Fortune to 0 in a game of $n = 8$ tosses; in particular, find the numerical value in the case of a balanced coin.

2. Repeat Exercise 1 for $n = 10$.

3. Find the general formula (in terms of n and p) for the expected number of returns of the Fortune to 0 in a game of $2n$ tosses.

4. Let X_1 and X_2 be indicator random variables. Show that they are independent if and only if the events $\{X_1 = 1\}$ and $\{X_2 = 1\}$ are independent.

✗ **5.** Let X_1 and X_2 be indicator random variables, and $Z = X_1 + X_2$. Show that Z is an indicator random variable if the events $\{X_1 = 1\}$ and $\{X_2 = 1\}$ are disjoint.

6. (Matching problem) Each of n persons leaves his hat in a checkroom. Suppose that the checks on the hats are "shuffled at random": each has probability $1/n$ of being returned to its proper owner. What is the expected number of persons who actually get back their own hats? (Hint: The Number of hats returning to their owners is a sum of n indicators.)

✗ **7.** (Birthday problem) What is the expected number of double birthdays (two or more persons having the same birthday) in a group of n persons? Here are the steps in the solution:

a) We suppose that each date of the year has equal probability of being the birthday of any person; hence, there is probability $\frac{1}{365}$ that a person is born on January 1. (We are not considering February 29 as a possible birthday; this hardly affects the results.) Show that the Number of persons in a group of n born on January 1 has the binomial distribution with $p = \frac{1}{365}$. (Assume that the birthdays are without "mutual effects"; for example, the members of the group are not siblings.)

b) Assuming the result in (a), find the probability that at least two persons in a group of n are born on January 1.

c) Let I_1 be the indicator random variable equal to 1 if at least two persons are born on January 1, and equal to 0 if not. Find $E(I_1)$.

d) Let $I_1, I_2, \ldots, I_{365}$ be the indicators corresponding to the successive days of the year, defined by analogy to I_1 in (c).

Show that the Number of double birthdays is equal to the sum of all these indicators; then find the expected value by applying Proposition 11.4.

8. Show that the variance of an indicator random variable X is $\Pr(X = 0) \times \Pr(X = 1)$.

11.3 PRODUCTS OF RANDOM VARIABLES

Definition 11.4 Let $[X, Y]$ be a composite random variable on a random trial. The *product* of X and Y, denoted $W = XY$, is defined as the random variable which associates with each outcome of the random trial the product of the values assigned by X and Y, respectively: if x and y are the values associated with a particular outcome by X and Y, respectively, then xy is the value associated with it by the random variable W.

Example 11.8 A coin is tossed four times. Let X and Y be the Numbers of H's appearing on the first two and last two tosses, respectively. Put $W = XY$; then the set of values of W is the set of integers arising as products of pairs of numbers selected from the set 0, 1, and 2: these are 0, 1, 2, and 4. An outcome in the event $\{X = 1, Y = 2\}$ is assigned the value $1 \cdot 2 = 2$ by the random variable W; an outcome in the event $\{X = 0, Y = 1\}$ is assigned the value $0 \cdot 1 = 0$, etc. ◁

Example 11.9 Let X and Y be the Scores on two tosses of a die; then $W = XY$ is the product of the Scores. It is a random variable with the set of values consisting of integers which are products of pairs of integers selected from the integers $1, \ldots, 6$.   ◁

The following proposition is analogous to Proposition 11.1.

Proposition 11.7 *The probability distribution of* $W = XY$ *can be calculated from the joint probability distribution of* $[X, Y]$: *for any value w of W, the probability* $\Pr(W = w)$ *is equal to the sum of all probabilities* $\Pr(X = x, Y = y)$ *for all values* $[x, y]$ *of* $[X, Y]$ *satisfying the equation* $w = xy$.

PROOF. The proof is exactly that of Proposition 11.1 except for one point: the equations of the form $z = x + y$ have to be replaced by equations of the form $w = xy$, and Z by W.   ◀

Example 11.10 Let X and Y be the Scores on two tosses of a die, and W the product of the Scores; then

$$\Pr(W = 1) = \Pr(X = 1, Y = 1) = \tfrac{1}{36},$$
$$\Pr(W = 2) = \Pr(X = 1, Y = 2) + \Pr(X = 2, Y = 1) = \tfrac{2}{36},$$
$$\Pr(W = 3) = \Pr(X = 1, Y = 3) + \Pr(X = 3, Y = 1) = \tfrac{2}{36},$$
$$\vdots$$
$$\Pr(W = 36) = \Pr(X = 6, Y = 6) = \tfrac{1}{36}.$$ ◁

A product of indicator random variables (Definition 11.3) is itself an indicator random variable; indeed, if each of the former has the set of values 0 and 1, then the products of the values consist just of the integers 0 and 1. If X and Y are indicator random variables and if W is their product, then the event $\{W = 1\}$ is identical with the intersection of the events $\{X = 1\}$ and $\{Y = 1\}$; in fact, an outcome is assigned the value 1 by the random variable W if and only if it is also assigned that value by both X and Y.

Example 11.11 A die is tossed twice. Let X be the indicator random variable which assigns the value 1 to the outcomes with even-numbered scores (2, 4, 6) and the value 0 to outcomes with odd-numbered scores (1, 3, 5) on the *first* toss; and let Y be the corresponding indicator random variable for the second toss. Here are some of the values of $[X, Y]$ associated with the outcomes of the pair of tosses:

outcome	X-value	Y-value
[1, 1]	0	0
[1, 2]	0	1
[1, 3]	0	0
[2, 1]	1	0
[2, 2]	1	1
[2, 3]	1	0

The events $\{X = 1\}$ and $\{Y = 1\}$ are the events that even-numbered scores appear on the first and second tosses, respectively. The product $W = XY$ is the random variable which assigns the numerical value 1 to outcomes of the two tosses which yield even-numbered scores on *both* tosses; therefore, the event $\{W = 1\}$ is the intersection of the two events $\{X = 1\}$ and $\{Y = 1\}$.   ◁

The relation between the expected value of the *sum* of two random variables and the expected values of the *summands* has been described by Proposition 11.2. A similar and simple relation for *products* of random variables is true under the additional assumption that the factor random variables are *independent*.

Proposition 11.8 *Let $[X, Y]$ be a composite random variable on a random trial such that X and Y are independent, and put $W = XY$; then $E(W)$ is equal to the product $E(X) \cdot E(Y)$; in other words, the expected value of the product is equal to the product of the expected values.*

PROOF. Before giving the proof of the general case, we note the very simple proof in the particular case in which X and Y are independent indicator random variables. By Proposition 11.5, the expected values of X and Y are $\Pr(X = 1)$ and $\Pr(Y = 1)$, respectively. The product W is also an indicator random variable and, by the same proposition, $E(W)$ is equal to $\Pr(W = 1)$. Now the event $\{W = 1\}$ is the intersection of the two events $\{X = 1\}$ and $\{Y = 1\}$; hence, by the independence of X and Y (Definition 9.1), the probability $\Pr(W = 1)$, which is the same as $\Pr(X = 1, Y = 1)$, is equal to the product $\Pr(X = 1)\Pr(Y = 1)$. Now we prove the general case.

1. $E(W)$ is the sum of all terms $w \cdot \Pr(W = w)$.

Reason. Definition 7.4.

2. The sum in Statement 1 is equal to the sum of all terms

$$xy \cdot \Pr(X = x, Y = y).$$

Reason. Every value w of W is of the form $x \cdot y$, by Definition 11.4; and, by Proposition 11.7, every probability $\Pr(W = w)$ is the sum of probabilities $\Pr(X = x, Y = y)$, where $w = xy$.

3. The sum in Statement 2 is the sum of all terms of the form

$$xy \cdot \Pr(X = x)\Pr(Y = y).$$

Reason. The relation $\Pr(X = x, Y = y) = \Pr(X = x)\Pr(Y = y)$ holds because X and Y are independent (Definition 9.1).

4. The sum in Statement 3 is the product of two sums: the sum of all terms $x \cdot \Pr(X = x)$ and the sum of all terms $y \cdot \Pr(Y = y)$.

Reason. The reason is similar to that for Statement 5 of the proof of Proposition 11.2. Let $x_1, x_2, \ldots$ and $y_1, y_2, \ldots$ be the sets of values of X and Y, respectively. The sum in the Statement may be written as

$$x_1 y_1 \cdot \Pr(X = x_1)\Pr(Y = y_1) + x_1 y_2 \cdot \Pr(X = x_1)\Pr(Y = y_2) + \cdots$$
$$+ x_2 y_1 \cdot \Pr(X = x_2)\Pr(Y = y_1) + x_2 y_2 \cdot \Pr(X = x_2)\Pr(Y = y_2) + \cdots$$
$$+ \cdots .$$

When the common factor $x_1 \cdot \Pr(X = x_1)$ is removed from each term in the first row, the sum of the remaining terms is

$$y_1 \cdot \Pr(Y = y_1) + y_2 \cdot \Pr(Y = y_2) + \cdots = E(Y);$$

and when the common factor $x_2 \cdot \Pr(X = x_2)$ is removed from each term in the second row, the sum of the remaining terms is also $E(Y)$. The same is true for every row. It follows that the sum of the *row sums* is $E(Y)$ multiplied by $x_1 \cdot \Pr(X = x_1) + x_2 \cdot \Pr(X = x_2) + \cdots = E(X)$. ◀

All the results given for products of two (independent) random variables can be generalized to products of two or more (independent) random variables.

Definition 11.5 Let $[X_1, \ldots, X_k]$ be a composite random variable on a random trial. The product of the random variables $X_1, \ldots, X_k$, denoted $W = X_1 \cdots X_k$, is defined as the random variable which associates with each outcome of the random trial the product of the values assigned by $X_1, \ldots, X_k$, respectively: if $[x_1, \ldots, x_k]$ is the value associated with an outcome by the composite random variable $[X_1, \ldots, X_k]$, then $w = x_1 \cdots x_k$ is the value associated with it by the random variable W.

Proposition 11.9 *The probability distribution of* $W = X_1 \cdots X_k$ *can be calculated from the joint probability distribution of* $[X_1, \ldots, X_k]$: *for any value w of W, the probability* $\Pr(W = w)$ *is equal to the sum of all probabilities* $\Pr(X_1 = x_1, \ldots, X_k = x_k)$ *for all multiplets* $[x_1, \ldots, x_k]$ *whose components satisfy the equation* $w = x_1 \cdots x_k$.

Proposition 11.10 *Let* $X_1, \ldots, X_k$ *be independent random variables on a random trial, and let W be their product; then* $E(W) = E(X_1) \cdots E(X_k)$.

We close this section by noting that the product of several indicator random variables is also an indicator random variable. If $X_1, \ldots, X_k$ are such indicator random variables, and if W is their product, then the event $\{W = 1\}$ is the intersection of the events $\{X_1 = 1\}, \ldots, \{X_k = 1\}$.

11.3 EXERCISES

1. Let X and Y be independent random variables with the common probability distribution

value:	1	2	3
probability:	.4	.4	.2

Find the probability distribution of $W = XY$. Find $E(W)$ by two methods: first, directly from the probability distribution of W; and second, by application of Proposition 11.8.

2. Let X_1, X_2, and X_3 be independent random variables with the common probability distribution given in Exercise 1. Put $W = X_1 X_2 X_3$ and proceed as in the latter exercise.

3. Find the probability distribution of the product of the two random variables given in Exercise 3 of the first section.

4. Find the probability distribution of the product of the Scores on two tosses of a die. Find the expected value of the product by two different methods as in Exercise 1.

5. Prove: If X and Y are indicator random variables, and the events $\{X = 1\}$ and $\{Y = 1\}$ are disjoint, then the product XY is identically equal to 0.

6. Verify the associative, commutative, and distributive laws for the sum and product of random variables; for example (commutative law), show that if X and Y are random variables, then $X + Y$ is the same random variable as $Y + X$, and XY the same as YX.

7. Show that every random variable may be written as a sum of multiples of indicators, i.e., in the form $x_1 I_1 + \cdots + x_k I_k$, where the x's are numbers forming the set of values of the random variable and the I's are indicators.

8. Proposition 11.8 was proved first for the special case of indicators and then for the general case. Show, by using Exercises 5 and 7, that the general case can be deduced from the special one.

9. Using Propositions 11.2 and 11.8, prove that the variance of the sum of two independent random variables is the sum of the respective variances.

11.4 SUMS OF A RANDOM NUMBER OF RANDOM VARIABLES

Let $X_1, \ldots, X_n$ be random variables on a random trial with a common expected value: $\mu = E(X_1)$. The expected value of their sum $X_1 + \cdots + X_n$ is, by Proposition 11.4, equal to $\mu + \cdots + \mu = n\mu$. In this section we shall derive the formula for the expected value of the sum $X_1 + \cdots + X_n$ in the case where n is not a fixed integer but a random variable, that is, where its value is determined by the outcome of a random trial.

We start with an example. Consider a coin-tossing game played in accordance with the following rules. The coin is tossed twice. If no H's appear, the game is terminated with a score of 0. If one H appears, the coin is tossed twice more, and we denote by X_1 the Number of H's on the latter tosses; the game is then terminated with a Score of X_1. If two H's appear on the first two tosses, then the coin is tossed four times. We let X_1 and X_2 stand for the Number of H's on the second pair of tosses and third pair of tosses, respectively; then the Number of H's on all four tosses is the sum of the random variables X_1 and X_2, and the game is terminated with a score of $X_1 + X_2$. In this game the terminal score is the sum of a *random number* of random variables: it is 0 if no H's appear on the first two tosses, X_1 if one H appears, and $X_1 + X_2$ if two H's appear.

We shall denote by S the random variable representing the terminal score, and by N the Number of H's on the first two tosses. Let I_0, I_1, and I_2 be the following indicator random variables: I_0 assigns the numerical value 1 to the outcomes in the event $\{N = 0\}$ and the value 0 to the outcomes in the complement of $\{N = 0\}$; I_1 and I_2 assign the numerical value 1 to the outcomes in the events $\{N = 1\}$ and $\{N = 2\}$, respectively, and the value 0 to the outcomes in their respective complements. The terminal score S has the following representation as a sum of products of random variables:

$$S = 0 \cdot I_0 + X_1 I_1 + (X_1 + X_2) I_2$$
$$= X_1 I_1 + X_1 I_2 + X_2 I_2. \qquad (11.1)$$

(The reader should be able to verify the correctness of the above equation on the basis of the definitions of the above random variables and the definitions of their sums and products.)

Equation (11.1) is used to compute the expected value of S. By virtue of Proposition 11.4, we find

$$E(S) = E(X_1 I_1) + E(X_1 I_2) + E(X_2 I_2). \qquad (11.2)$$

It will be shown that X_1 and I_1, X_1 and I_2, and X_2 and I_2 are three pairs of *independent* random variables; assuming this to be true, and applying Proposition 11.8, we deduce that the right-hand side of Eq. (11.2) is equal to

$$E(X_1)E(I_1) + E(X_1)E(I_2) + E(X_2)E(I_2). \qquad (11.3)$$

It follows from Proposition 11.5 that $E(I_1) = \Pr(N = 1)$ and $E(I_2) = \Pr(N = 2)$; furthermore, X_1 and X_2 have the same distribution, so that $E(X_1) = E(X_2)$; hence, the sum in expression (11.3) is equal to

$E(X_1)[\Pr(N = 1) + 2 \cdot \Pr(N = 2)]$
$= E(X_1)[0 \cdot \Pr(N = 0) + 1 \cdot \Pr(N = 1) + 2 \cdot \Pr(N = 2)]$
$= E(X_1)E(N)$ *E value of n tosses of a coin is np.*   (11.4)
$= (2p)(2p) = 4p^2.$ *Since X_1 depends on 2 tosses*
$\qquad\qquad\qquad\qquad \therefore \ E(X_1) = 2p$

In order to complete the proof of the above chain of equations, we still have to show the independence of the pairs of random variables inside the expected value sign in Eq. (11.2). The proof of the independence of X_1 and I_1 is typical of the proofs for the other two pairs. The independence of X_1 and I_1 is characterized by the validity of the following equations:

$$\Pr(X_1 = 0, I_1 = 0) = \Pr(X_1 = 0)\Pr(I_1 = 0),$$
$$\Pr(X_1 = 1, I_1 = 0) = \Pr(X_1 = 1)\Pr(I_1 = 0),$$
$$\Pr(X_1 = 2, I_1 = 0) = \Pr(X_1 = 2)\Pr(I_1 = 0),$$
$$\Pr(X_1 = 0, I_1 = 1) = \Pr(X_1 = 0)\Pr(I_1 = 1),$$
$$\Pr(X_1 = 1, I_1 = 1) = \Pr(X_1 = 1)\Pr(I_1 = 1),$$
$$\Pr(X_1 = 2, I_1 = 1) = \Pr(X_1 = 2)\Pr(I_1 = 1).$$
$$(11.5)$$

Let us verify the fifth equation: $\Pr(X_1 = 1, I_1 = 1) = \Pr(X_1 = 1)\Pr(I_1 = 1)$; the others are verified in a similar way. The events $\{I_1 = 1\}$ and $\{X_1 = 1\}$ are unions of disjoint events determined by the first two and second two tosses, respectively: $\{I_1 = 1\}$ is the union of the disjoint events "H and T on the first and second tosses, respectively" and "T and H on the first and second tosses, respectively"; and $\{X_1 = 1\}$ is the union of similar events for the third and fourth tosses. The independence of the events $\{I_1 = 1\}$ and $\{X_1 = 1\}$ follows from the Independence Theorem (Section 5.3) and Proposition 5.7; hence, the fifth equation above is just the defining equation of independence.

The part of this example that will be generalized is contained in Eq. (11.4).

Proposition 11.11 *Let $X_1, \ldots, X_k$ and N be a set of $k + 1$ random variables on a random trial such that*

a) *$X_1, \ldots, X_k$ have a common expected value μ;*
b) *N has the set of values $0, 1, \ldots, k$;*
c) *N and $X_1, \ldots, N$ and X_k are pairs of independent random variables.*

Let the random variable S be defined as the sum of a random number N of X's: it assigns the numerical value 0 to outcomes in $\{N = 0\}$; it assigns to outcomes in $\{N = 1\}$ the same numerical values as those assigned by X_1; and, it assigns to outcomes in $\{N = j\}$ the same values as those assigned by the sum $X_1 + \cdots + X_j$ of the first j X's, $j = 1, \ldots, k$. The expected value of S is related to the expected values of the X's and N by the following simple equation:

$$E(S) = \mu E(N). \tag{11.6}$$

PROOF

1. Let $I_1, \ldots, I_k$ be indicator random variables defined as follows: I_1 assigns the value 1 to outcomes in the event $\{N = 1\}$ and 0 to all others, I_2 assigns the value 1 to outcomes in $\{N = 2\}$ and 0 to all others, $\ldots$. Then S is representable as a sum of products of random variables:

$$\begin{aligned}
S &= X_1 I_1 + (X_1 + X_2)I_2 + (X_1 + X_2 + X_3)I_3 \\
&\quad + \cdots + (X_1 + \cdots + X_k)I_k \\
&= X_1 I_1 + X_1 I_2 + \cdots + X_1 I_k \\
&\quad + X_2 I_2 + \cdots + X_2 I_k \\
&\quad\quad + \cdots + \\
&\quad\quad\quad + X_k I_k.
\end{aligned}$$

Reason. This is an immediate consequence of the definitions of S and of the sums and products (cf. Section 11.3, Exercise 6).

2. Each pair X, I in the products in the above expression for S is a pair of *independent* random variables; for example, X_1 and I_3 are independent.

Reason. We prove just the independence of X_1 and I_3 as it is typical of the other pairs. The event $\{I_3 = 1\}$ is identical with the event $\{N = 3\}$; hence, for each value x of X_1, the independence of X_1 and N implies the relations

$$\Pr(X_1 = x, I_3 = 1) = \Pr(X_1 = x, N = 3)$$
$$= \Pr(X_1 = x)\Pr(N = 3) \qquad (11.7)$$
$$= \Pr(X_1 = x)\Pr(I_3 = 1).$$

It remains for us to prove the validity of the equation

$$\Pr(X_1 = x, I_3 = 0) = \Pr(X_1 = x)\Pr(I_3 = 0). \qquad (11.8)$$

This is a consequence of Proposition 5.8, because $I_3 = 0$ is the complement of $I_3 = 1$.

3. $E(S) = E(X_1I_1) + E(X_1I_2) + \cdots + E(X_1I_k)$
$$+ E(X_2I_2) + \cdots + E(X_2I_k)$$
$$+ \cdots +$$
$$+ E(X_kI_k).$$

Reason. Apply Proposition 11.4 to the representation of S in Statement 1.

4. The expected values of the products on the right-hand side of the equation in Statement 3 are the products of the corresponding expected values: $E(X_1I_1) = E(X_1)E(I_1)$, $E(X_1I_2) = E(X_1)E(I_2)$, ...

Reason. Statement 2 and Proposition 11.8.

5. The sum on the right-hand side of the equation in Statement 3 is equal to

$$\mu[\Pr(N = 1) + 2 \cdot \Pr(N = 2) + 3 \cdot \Pr(N = 3) + \cdots + k \cdot \Pr(N = k)].$$

Reason. By Statement 4, the terms on the right-hand side are equal to

$$\mu \cdot \Pr(I_1 = 1), \qquad \mu \cdot \Pr(I_2 = 1), \qquad \ldots,$$

respectively, which are the same as

$$\mu \cdot \Pr(N = 1), \qquad \mu \cdot \Pr(N = 2), \qquad \ldots,$$

respectively; therefore, the sum may be written as

$$\mu \cdot \Pr(N = 1) + \mu \cdot \Pr(N = 2) + \cdots + \mu \cdot \Pr(N = k)$$
$$+ \mu \cdot \Pr(N = 2) + \cdots + \mu \cdot \Pr(N = k)$$
$$+ \cdots +$$
$$+ \mu \cdot \Pr(N = k).$$

Summation by *columns* yields the form indicated in the statement.

6. Equation (11.6) follows from Statement 5.

Reason. Definition of $E(N)$.   ◄

11.4 EXERCISES

1. Verify the second equation in the set in formula (11.5).

2. Give a *complete* justification for the representation of S in Statement 1 of the proof of Proposition 11.11.

3. Show that the proof of Proposition 11.11 implies the following generalization of formula (11.6) to the case where the X's have different expected values: If $\mu_1 = E(X_1), \ldots, \mu_k = E(X_k)$, then

$$E(S) = \mu_1 \cdot \Pr(N = 1) + (\mu_1 + \mu_2)\Pr(N = 2)$$
$$+ \cdots + (\mu_1 + \cdots + \mu_k)\Pr(N = k).$$

4. Let $X_1, \ldots, X_k, N$ be random variables on a random trial, satisfying the hypothesis of Proposition 11.11.

a) By similarity to the definition of S, define the *product* P of a random number of the X's.

b) Derive a representation of P similar to that of S in Statement 1 of the proof.

c) Assume now that the random variables $X_1, \ldots, X_k, N$ are independent. Prove the validity of the equation

$$E(P) = \mu \cdot \Pr(N = 1) + \mu^2 \cdot \Pr(N = 2) + \cdots + \mu^k \cdot \Pr(N = k).$$

11.5 APPLICATION TO BRANCHING PROCESSES

We describe a *branching process*. A single "particle" is given. It reproduces, generating a number of particles identical with it, forming the "first generation." The number of particles in the first generation is not known beforehand: it is assumed to be determined by the outcome of a random trial; that is, it is a random variable. Let us denote this random variable by X_1; its set of values is a set of integers 1, 2, ..., k representing the possible sizes of the first generation. The process is said to "die out" before the first generation if the number of particles in the first generation is 0; this is the event $\{X_1 = 0\}$. If the process does not die out before the first generation, a "second generation" is created in the following way: each particle in the first generation reproduces in accordance with the same law governing the reproduction of the original particle. The numbers of particles produced by the members of the first generation are random variables with a common probability distribution, that of X_1; furthermore, they are independent of X_1, the size of the first generation. The particles produced by the members of the first generation form the second generation; let X_2 be the random variable representing the number of particles in the second generation. If none of the first-generation particles produces any descendants, the process "dies out before the second generation"; otherwise, a third generation is born. The general rule of passage from one generation to the next is as follows. Let X_n be the size of the nth generation; if X_n is not 0, then each particle in the nth generation reproduces in accordance with the same probability distribution as X_1, the first generation; furthermore, the random variables representing the numbers of descendants produced by the respective members of the nth generation are independent of X_n, the number of "parents." A consequence of this description is the following assertion:

Let X_n be the size of the nth generation and Y_1, Y_2, ... the numbers of particles generated by the respective members of the nth generation. Let X_{n+1} be the size the $(n + 1)$st generation. The random variables X_n and Y_1, X_n and Y_2, ... are respectively independent, and X_{n+1} is the sum of a random number X_n of random variables Y_1, Y_2, ...

This is a mathematical idealization of processes in the real world; here are a few examples.

1) *Spread of infectious disease.* A carrier of a contagious disease enters a community of healthy people; he is the first "particle" in the branching process. He infects a certain number of persons; these form the "first generation." Each of these infects a number of other people, and so on. It must be mentioned that our mathematical model is an idealization of a very much simpler process than the spread of disease that *actually* takes place. In the mathematical model, we assumed that the members of later generations infect others in the same way as the original carrier. This is not entirely realistic: as disease spreads through a community, more people

gain immunity by infection, and the number of new cases gradually goes down (cf. Exercise 2, below).

2) *Survival of family names.* The family name is transmitted from generation to generation through the male members. The males of each generation are the "particles"; their male offspring are the particles produced by them. The family name dies out as soon as no males are born in some generation.

3) *Microorganic growth.* Some cells in plants and animals reproduce by a process of binary fission. A single cell splits in two, forming two cells identical with it; each of these splits in turn, and so on. One new cell is created each time fission occurs; on the other hand, some of the living cells die. The number of fissions and deaths occurring in a given unit of time varies with the conditions of the cells. If the "colony" of cells is observed at regular epochs of time, the numbers of cells observed represent the sizes of successive generations.

Now we apply Proposition 11.11 to finding the expected size of the successive generations of a branching process.

Proposition 11.12 *Let* $X_1, X_2, \ldots, X_n, \ldots$ *be the successive sizes of the generations of a branching process; and let* $\mu = E(X_1)$. *Then*

$$E(X_n) = \mu^n, \qquad n = 1, 2, \ldots ; \qquad (11.9)$$

in other words, the expected sizes of successive generations are the powers of the expected size of the first generation.

PROOF

1. Let $Y_1, Y_2, \ldots$ be the random variables representing the numbers of particles produced by the members of the $(n - 1)$st generation. Then the random variables $Y_1, Y_2, \ldots$, and X_{n-1} (the size of the $(n - 1)$st generation) satisfy the same conditions as the random variables $X_1, \ldots, X_k$ and N, respectively, in the hypothesis of Proposition 11.11; furthermore, X_n, the size of the nth generation, is the sum of a random number X_{n-1} of Y's.

Reason. This was assumed in the above mathematical model of the branching process.

2. $E(Y_1) = \mu.$

Reason. The number of particles produced by any member of any generation has the same probability distribution as X_1; hence, it has the same expected value μ.

3. $E(X_n) = \mu E(X_{n-1})$.

Reason. Let X_n and X_{n-1} take the place of S and N in Eq. (11.6).

4. Equation (11.9) follows from Statement 3.

Reason. Statement 3 is valid for all n larger than 1; hence, $E(X_2) = \mu E(X_1) = \mu^2$ (for $n = 2$), $E(X_3) = \mu E(X_2) = \mu \cdot \mu^2 = \mu^3$ (for $n = 3$), and so on. ◄

Interesting implications may be derived from this proposition. If μ, the expected number of particles produced by a single particle is equal to 1, then $E(X_n) = 1^n = 1$ for all values of n. If μ is larger than 1, the expected size of the nth generation, μ^n, increases rapidly with n. Finally, suppose that μ is less than 1. The expected size of the nth generation, μ^n, rapidly gets smaller as the number of generations increases; furthermore, the expected *total number* of particles ever produced never exceeds the quantity $\mu/(1 - \mu)$ no matter how long the process continues. This is proved as follows. The total number of particles in the first n generations is the sum of random variables $X_1 + \cdots + X_n$. By Propositions 11.4 and 11.12, we have

$$E(X_1 + \cdots + X_n) = E(X_1) + \cdots + E(X_n) = \mu + \mu^2 + \cdots + \mu^n.$$

By Proposition 10.3 (with μ instead of q), the last sum is equal to

$$\mu/(1 - \mu) - \mu^{n+1}/(1 - \mu),$$

which is certainly less than $\mu/(1 - \mu)$.

11.5 EXERCISES

1. Generalize Eq. (11.9) in Proposition 11.2 to the case where the expected number of births changes from one generation to the next. Let μ_1 be the expected number of members of the first generation, and μ_n ($n > 1$) the expected numbers of *descendants* of each member of the $(n - 1)$st generation. Prove:

$$E(X_n) = \mu_1 \cdot \mu_2 \cdots \mu_n.$$

2. How can the result of Exercise 1 be used to make the branching-process model more applicable to the problem of the spread of infectious disease?

3. Suppose that the sequence of numbers $\mu_1, \mu_2, \ldots, \mu_n, \ldots$ in Exercise 1 has a limit μ. Discuss the behavior of $E(X_n)$ as n increases in the cases: (i) $\mu < 1$, (ii) $\mu > 1$.

chapter 12

THE GENERAL LAW OF LARGE NUMBERS, CONDITIONAL PROBABILITY, AND MARKOV CHAINS

This chapter consists of three loosely linked subjects. The first, the General Law of Large Numbers, is an extension of the Law from the context of coin tossing to more general random variables. The second, conditional probability, is defined in the framework of coin tossing. The General Law of Large Numbers illuminates the meaning of conditional probability, giving it a long-run frequency interpretation. The last topic, Markov chains, is treated only in the particular case of the alternate tossing of each of two coins. An "ergodic" theorem is proved by simple algebraic methods.

12.1 THE GENERAL LAW OF LARGE NUMBERS

In the analysis of numerical measurements it is customary to compute averages and use them as summary descriptions of groups of data. There is a theoretical justification for this procedure: if a group of measurements is considered to be a set of independent random variables with a common probability distribution, then the average of a large number of these measurements will, with probability close to 1, be near to the common expected value. This phenomenon was explained in the particular case of the tossing of a die (Section 9.4): the average of many Scores is, with high probability, close to the expected value of the Score on one toss, namely, $3\frac{1}{2}$. The proof of the general case is based on the estimates used in the proof of the Clustering Principle.

Let $[X_1, \ldots, X_n]$ be a composite random variable on a random trial. The random variable $X_1 + \cdots + X_n$ was defined in Definition 11.2; now we define the *average* of the X's,

$$(X_1 + \cdots + X_n)/n,$$

as that random variable which associates with each outcome of the random trial the average of the values associated by $X_1, \ldots, X_n$, respectively: if $[x_1, \ldots, x_n]$ is a multiplet associated with an outcome by $[X_1, \ldots, X_n]$, then $(x_1 + \cdots + x_n)/n$ is the value associated with it by the random variable $(X_1 + \cdots + X_n)/n$. The following proposition reveals the significance of the expected value of a random variable: the average of a large number of independent random variables tends, with high probability, to be close to their common expected value.

General Law of Large Numbers. *Let $X_1, \ldots, X_n$ be independent random variables on a random trial with a common probability distribution; and let $E(X_1)$ be their common expected value. For any arbitrary positive number d, let P_n be the probability that the average of the X's,*

$$(X_1 + \cdots + X_n)/n,$$

lies between $E(X_1) - d$ and $E(X_1) + d$. The conclusion is: the limit of the sequence of probabilities $P_1, P_2, \ldots, P_n, \ldots$ is 1.

PROOF. Let $x_1, \ldots, x_k$ be the common set of values of $X_1, \ldots, X_n$. We exclude the trivial case in which there is only a single value.

1. The average $(X_1 + \cdots + X_n)/n$ is equal to the weighted sum

$$x_1 f_1 + \cdots + x_k f_k, \tag{12.1}$$

where $f_1, \ldots, f_k$ are the relative frequencies of $x_1, \ldots, x_k$.

Reason. The sum $X_1 + \cdots + X_n$ is equal to the sum of the terms

$$x_1 \cdot \text{number of } X\text{'s equal to } x_1,$$
$$x_2 \cdot \text{number of } X\text{'s equal to } x_2,$$
$$\vdots$$
$$x_k \cdot \text{number of } X\text{'s equal to } x_k;$$

hence, the average of the X's is obtained upon division of each term by n.

Pr$(|A_n - E(X)| < d)$ $\geq 1 - \dfrac{K}{4nc^2}$

2. Let $A_1, \ldots, A_k$ be the events

$A_1 = $ "f_1 is between $\Pr(X_1 = x_1) - c$ and $\Pr(X_1 = x_1) + c$,"

$A_2 = $ "f_2 is between $\Pr(X_1 = x_2) - c$ and $\Pr(X_1 = x_2) + c$,"

$\vdots$

$A_k = $ "f_k is between $\Pr(X_1 = x_k) - c$ and $\Pr(X_1 = x_k) + c$";

then, the probability of their intersection is at least equal to

$$1 - \frac{k}{4nc^2}.$$

Reason. Proposition 9.5 (with c in place of d).

3. The difference between the weighted sum (12.1) and $E(X_1)$ may be written as the sum

$$x_1\big(f_1 - \Pr(X_1 = x_1)\big) + \cdots + x_k\big(f_k - \Pr(X_1 = x_k)\big).$$

Reason. Definition of $E(X_1)$ and factorization of $x_1, \ldots, x_k$.

4. The absolute value of the sum displayed in Statement 3 is at most equal to $c|x_1| + \cdots + c|x_k|$ if the coefficients of the x's are all smaller in absolute value than c.

Reason. We use the property of the real numbers that the absolute value of a sum is at most equal to the sum of the absolute values of the summands.

5. Define d as $c|x_1| + \cdots + c|x_k|$; then the intersection of the events $A_1, \ldots, A_k$ implies the event "the sum (12.1) is between $E(X_1) - d$ and $E(X_1) + d$."

Reason. Statement 4 and the meaning of "implies" (Definition 5.6).

6. The probability that the average of the X's is between $E(X_1) - d$ and $E(X_1) + d$ is at least equal to $1 - k/4nc^2$.

Reason. Statements 1, 2, and 5 and Proposition 5.4.

7. The conclusion of the General Law now follows.

Reason. By the exclusion of the trivial case mentioned above, at least one of the x-values is not 0; thus, for any positive number d, there *is* a positive number c defined as $c = d/(|x_1| + \cdots + |x_k|)$; therefore, by Statement 6, the sequence of probabilities has the limit 1. ◀

Example 12.1 We illustrate the estimates used above in the case of n tosses of a die. The set of values for the Score is the set of integers 1 through 6. For an arbitrary positive number d, let c be the number

$$d/(1 + 2 + 3 + 4 + 5 + 6) = d/21;$$

here, $k = 6$. By Statement 6 above, the probability that the average of the Scores of n tosses of a die differs from 3.5 by at most d units is at least equal to

$$1 - \frac{k}{4nc^2} = 1 - \frac{6}{4n(d/21)^2} = 1 - \frac{661\frac{1}{2}}{nd^2}. \qquad \triangleleft$$

12.1 EXERCISES

1. Consider the 25 observed values of the random variables in Exercise 1, Section 9.5. For this particular case verify Statement 1 of the above proof.

2. Refer to Exercise 2, Section 9.5. What is the observed average number of black balls in the Sample in 25 draws? Compare this to the expected number in the Sample.

3. Formulate a statement analogous to that in Example 12.1 for n tosses of a *pair* of dice and for the average of the sums of the *two* Scores.

4. Why is the proof of the General Law of Large Numbers valid only for independent random variables?

5. Let the independent random variables $X_1, \ldots, X_n$ have the common probability distribution:

outcome:	-1	0	$+1$
probability:	.2	.2	.6

Estimate the probability that the average of the X's differs from .4 by at most d units, as in Example 12.1.

12.2 CONDITIONAL PROBABILITY

The law of large numbers for coin tossing (Chapter 3) is a particular case of the general law derived in the previous section; this can be verified with the help of indicator random variables (Definition 11.3). Let $[X_1, \ldots, X_n]$ be the following composite random variable on the random trial consisting of n tosses of a coin: X_1 assigns the numerical value 1 to outcomes in the event "H on the first toss" and the value 0 to its complement ("T on the first toss"); X_2 assigns the numerical value 1 to outcomes in the event "H on the second toss" and 0 to its complement; and so on. The X's are indicator random variables. The sum $X_1 + \cdots + X_n$ is the Number of H's, and the average $(X_1 + \cdots + X_n)/n$ is the ratio of the Number of H's to the number of tosses. The X's have a common probability distribution:

$$Pr(X_1 = 1) = Pr(X_2 = 1) = \cdots = p,$$
$$Pr(X_1 = 0) = Pr(X_2 = 0) = \cdots = q.$$

The X's are also independent: the events $\{X_1 = 1\}$, $\{X_2 = 1\}$, ... are determined by different tosses, and hence, by the Independence Theorem, are independent; therefore,

$$Pr(X_1 = 1, X_2 = 1, \ldots) = Pr(X_1 = 1) \cdot Pr(X_2 = 1) \cdots .$$

The same reasoning applies to all the other probabilities in the joint probability distribution of $[X_1, \ldots, X_n]$; for example,

$$Pr(X_1 = 1, X_2 = 0, \ldots) = Pr(X_1 = 1) \cdot Pr(X_2 = 0) \cdots .$$

The expected value of each of the X's is $Pr(X_1 = 1) = p$, by Proposition 11.5. The General Law of Large Numbers applies to the random variables $X_1, \ldots, X_n$: their average—which is the proportion of H's—tends to be close to p when n is large. This is exactly the law of large numbers for coin tossing.

Consider the following modification of the coin game, played with several coins. There are three coins with probabilities p_0, p_1, and p_2 of H, respectively. The coin with probability p_0 is tossed first: if H appears, then the coin with probability p_1 is tossed; if T appears, then the coin with probability p_2 is tossed. Let us refer to the coins with probabilities p_1 and p_2 as coins I and II, respectively; p_0 and $1 - p_0 = q_0$ are called the *a priori* probabilities of coins I and II, respectively. The system of possible outcomes of the two tosses is [H H], [H T], [T H], [T T]; the corresponding probabilities are $p_0 p_1$, $p_0 q_1$, $q_0 p_2$, and $q_0 q_2$, respectively, where $q_1 = 1 - p_1$ and $q_2 = 1 - p_2$. The event "H on the second toss" consists of the two outcomes [H H] and [T H] and so has probability $p_0 p_1 + q_0 p_2$; and the event "T on the second toss" consists of the two outcomes [H T] and [T T] and so has probability $p_0 q_1 + q_0 q_2$. Now we define *conditional probability*.

Definition 12.1 The conditional probability of coin I given H on the second toss is

$$p_0 p_1 / (p_0 p_1 + q_0 p_2);$$

the conditional probability of coin I given T on the second toss is

$$p_0 q_1 / (p_0 q_1 + q_0 q_2).$$

The corresponding conditional probabilities for coin II are obtained from these by interchanging q_0, p_2, and q_2 with p_0, p_1, and q_1, respectively, in the numerators above.

These conditional probabilities are also known as *a posteriori* probabilities; the corresponding formulas are special cases of a general formula known as "Bayes's Theorem." The conditional probabilities have the characteristic properties of ordinary probabilities in a distribution; they are nonnegative and their sum is 1:

conditional probability of coin I, given H

$+$ conditional probability of coin II, given H

$$= \frac{p_0 p_1}{p_0 p_1 + q_0 p_2} + \frac{q_0 p_2}{q_0 p_2 + p_0 p_1} = 1.$$

The conditional probabilities given T also have the sum 1.

We now describe the *operational* meaning of conditional probability. Suppose the pair of tosses just described is repeated several times under identical conditions. Consider the two proportions: the proportion of pairs in which coin I is selected on the first toss and in which H appears on the second toss; and the proportion of pairs in which H appears on the second toss. The ratio of the proportions (the first proportion divided by the second) is

$$\frac{\text{number of pairs in which coin I is selected and H appears on the second toss}}{\text{number of pairs in which H appears on the second toss}}.$$

$$(12.2)$$

It will be shown, as a consequence of the General Law of Large Numbers, that this ratio is, with high probability, very close to the conditional probability of coin I given H on the second toss (Definition 12.1). Similar results hold for the other appropriate ratios.

The proof of the above assertion is based on the proof of the law of large numbers for coin tossing given in the opening paragraph of this section.

Suppose there are n pairs of tosses, performed identically as the pair described above: the coin with probability p_0 is tossed first, and then coin I or coin II is selected in accordance with the Outcome of the first toss, and is tossed. Let $X_1, \ldots, X_n$ be indicator random variables defined as follows: X_1 assigns the numerical value 1 to outcomes in the event "coin I is selected and H appears on the second toss of the first pair of tosses," and the value 0 to all other outcomes; and $X_2, \ldots, X_n$ are analogously defined for the second, ..., nth pairs of tosses, respectively. It is assumed that $X_1, \ldots, X_n$ are independent random variables, in accordance with the empirical assumption that the n pairs of tosses are performed without mutual effects among the Outcomes. The common expected value of the X's is the probability of the event "coin I is selected and H appears on the second toss"; thus, by hypothesis, it is equal to $p_0 p_1$. It follows that the average of the X's,

$$(X_1 + \cdots + X_n)/n,$$

is, with high probability, close to $p_0 p_1$ when n is very large; in other words, the proportion corresponding to the numerator in (12.2) is very close to $p_0 p_1$. Now let $Y_1, \ldots, Y_n$ be indicator random variables defined as follows: Y_1 assigns the numerical value 1 to outcomes in the event "H appears on the second of the first pair of tosses" and the value 0 to the outcomes in the complement; and $Y_2, \ldots, Y_n$ are analogously defined for the second, ..., nth pairs of tosses, respectively. The Y's are also independent, with the common expected value equal to the probability of the event "H appears on the second of the first pair of tosses," and hence, equal to $p_0 p_1 + q_0 p_2$. The average of the Y's is the proportion corresponding to the denominator in (12.2). By the General Law of Large Numbers, this average is, with high probability, close to $p_0 p_1 + q_0 p_2$ when n is large. It follows that the ratio (12.2) is, with high probability, close to the ratio defining the conditional probability of coin I given H (Definition 12.1).

Example 12.2 Suppose that inoculation furnishes partial protection against a specific disease: we assume that a person with inoculation has probability p_1 of getting the disease, and that a person without it has probability p_2, where p_1 is less than p_2. Suppose that the proportion of inoculated persons in a population is p_0; what is the conditional probability that a person has been inoculated, given that he has the disease?

The health of an individual can be thought of as being determined by two tosses of two coins. There are three coins with probabilities p_0, p_1, and p_2 of H, respectively. The coin with probability p_0 is tossed first: if H appears, the individual is inoculated; if T appears, he is not. In the former case the coin with probability p_1 is tossed: if H appears, he contracts the disease, and if T appears, he does not. In the latter case, the coin with probability p_2 is tossed: he contracts the disease or does not accordingly as H or T, respectively, turns up. The conditional probability that a person has been inoculated, given that he has the disease, is interpreted as the conditional probability of coin I given H on the second toss, and is given by Definition 12.1.

Here is a numerical illustration. Suppose that 75 percent of the population is inoculated, and that the probabilities of contraction of the disease are .4 and .8 for the inoculated and uninoculated groups, respectively. The conditional probabilities are:

cond. prob. of inoculated, given disease
$$= (.75)(.4)/[(.75)(.4) + (.25)(.8)] = .6,$$
cond. prob. of not inoculated, given disease
$$= (.25)(.8)/[(.75)(.4) + (.25)(.8)] = .4,$$
cond. prob. of inoculated, given no disease
$$= (.75)(.6)/[(.75)(.6) + (.25)(.2)] = .9,$$
cond. prob. of not inoculated, given no disease
$$= (.25)(.2)/[(.75)(.6) + (.25)(.2)] = .1.$$

The operational meaning of these conditional probabilities is deduced from the previous discussion; for example, the conditional probability that a person has been inoculated, given that he has the disease, is approximately equal to the proportion of the inoculated persons among those having the disease. Though the proportion seems to be high (.6), it is much lower than the proportion of inoculated persons among those free of the disease (.9). ◁

Example 12.3 Suppose that an electorate is divided into two parties: a proportion p_0 belongs to the Blue party and a proportion $q_0 = 1 - p_0$ to the Green party. Suppose that a proportion p_1 of the Blue party is in favor of a given public policy (and the remaining proportion $q_1 = 1 - p_1$ against it), and a proportion p_2 of the Green party is in favor (and the remaining proportion $q_2 = 1 - p_2$ opposed). A voter is selected at random from the electorate; what is the conditional probability that he is a Blue party member, given that he is in favor of the policy?

This problem is analogous to the previous one involving the inoculations. The Blue and Green parties are like the "inoculated" and "uninoculated" groups, respectively; and those in favor and those against the policy are analogous to the "diseased" and "healthy" groups, respectively.   ◁

12.2 EXERCISES

1. Sixty percent of a population is inoculated. The proportions of the inoculated and uninoculated groups having a disease are .1 and .9, respectively. Find the proportion of the diseased group which is not inoculated; and the proportion of the healthy group.

2. Repeat Exercise 1 with the changes: 50 percent, .2, .7.

3. Criticize the following statement: "Most people having a disease are found to have been inoculated; therefore, an uninoculated person is less likely to have the disease than one who is inoculated."

4. Construct a statement similar to the one quoted in Exercise 3 for parties and policies.

12.3 MARKOV CHAINS

An important use of a related concept of conditional probability is in the theory of Markov chains, which we now present in its most rudimentary form. Suppose a player has two coins with probabilities p_1 and p_2 of H, respectively; let us call these coin I and coin II, respectively. He begins by tossing one of the coins: if H appears, he tosses it again; if T appears, he switches to the other coin and tosses it. The player stays with a coin as long as H keeps appearing, switching to the other coin as soon as T appears, and then continuing in the same way with the latter coin. We call p_1 the "conditional probability of staying with coin I, given coin I," and p_2 the "conditional probability of staying with coin II, given coin II." The probabilities $q_1 = 1 - p_1$ and $q_2 = 1 - p_2$ are the respective conditional probabilities of switching, given coins I and II, respectively.

This game is an idealization of a "system" with two "states" in which the system alternately stays in one state, moves to the other, then back, etc. The game can be generalized to represent a system with any number s of states. Consider s urns, labeled $1, 2, \ldots, s$, respectively, each containing several balls of each of s different colors, which are also labeled $1, 2, \ldots, s$. We draw a ball at random from one of the urns, noting the color of the ball drawn: we replace the ball and then move to the urn with the same number as the color of the ball just drawn. The same procedure is repeated for the second urn, and so on. This system is called a Markov chain with s states. We shall study just the game with the two coins, a Markov chain with two states.

Now we formally define a system of outcomes and probabilities for the two-coin game of n tosses. The system of outcomes is, as in Definition 2.1, the system of all multiplets of n letters H and T: H or T appears in the kth position in the multiplet accordingly as H or T appears on the kth toss. The assignment of probabilities is different from that in the game with one coin (Definition 2.2). The probability of an outcome is a product of n factors p_1, q_1, p_2, and q_2: there is a factor p_1 (or q_1) for each H (or T) appearing on a toss of coin I, and a factor p_2 (or q_2) for each H (or T) appearing on a toss of coin II; thus, in contrast to the one-coin game, the probability of a multiplet depends not only on the number of letters H in it, but also on the order in which they occur and also on which coin is used first.

The Consistency Theorem (Section 5.2) holds for this assignment of probabilities: the probability of an event determined by the first k tosses in a game of n tosses, $k < n$, is the same as for a game of k tosses. The proof given above for the one-coin game can be suitably modified to cover the two-coin game: the *two* equations $p_1 + q_1 = 1$ and $p_2 + q_2 = 1$ are used instead of a single equation. The details of the proof are left to Exercise 5 below, where the outline is given. The Consistency Theorem implies—but does not depend on the fact—that the sum of the probabilities of all the outcomes is 1; indeed, the events "H on the first toss" and "T on the first toss" are disjoint, the sum of their probabilities is 1, and every outcome belongs to either one event or the other.

Example 12.4 Let us illustrate this assignment of probabilities for a game of $n = 3$ tosses in the case where coin I is tossed first:

outcome	probability	outcome	probability
[HHH]	$p_1 p_1 p_1$	[HTT]	$p_1 q_1 q_2$
[HHT]	$p_1 p_1 q_1$	[THT]	$q_1 p_2 q_2$
[HTH]	$p_1 q_1 p_2$	[TTH]	$q_1 q_2 p_1$
[THH]	$q_1 p_2 p_2$	[TTT]	$q_1 q_2 q_1$

Using the equations $p_1 + q_1 = p_2 + q_2 = 1$, one verifies that the sum of the probabilities is 1. The system of probabilities for such a game begun with coin II is obtained from the one above by interchanging the subscripts 1 and 2; for example, the probability of [HTT] is $p_2 q_2 q_1$. ◁

Until indicated otherwise, we shall suppose that the first toss is performed with coin I. Let $E_1, \ldots, E_n$ be the events

E_1 = "H appears on the first toss,"

E_k = "an even number of T's appear on the first k tosses,"

$\vdots$

E_n = "an even number of T's appear on the n tosses,"

and $F_1, \ldots, F_n$ their respective complements. Let $u_1, \ldots, u_n$ be the probabilities of $E_1, \ldots, E_n$, and $v_1, \ldots, v_n$ those of $F_1, \ldots, F_n$; thus,

$$u_1 + v_1 = 1, \quad \ldots, \quad u_n + v_n = 1.$$

Since coin I is the first to be tossed, the event E_{k-1} is equivalent to "coin I is tossed on the kth toss," and F_{k-1} to the corresponding event for coin II.

Example 12.5 For the game of three tosses in Example 12.4, the probabilities u_1, u_2, and u_3 are:

$$u_1 = p_1, \qquad u_2 = p_1^2 + q_1 q_2,$$
$$u_3 = p_1^3 + p_1 q_1 q_2 + q_1 p_2 q_2 + q_1 q_2 p_1. \qquad \triangleleft$$

An important relation among all the probabilities u and v is contained in the following proposition.

Proposition 12.1 *The following equations hold:*

$$u_1 = p_1,$$
$$u_2 = u_1 p_1 + v_1 q_2,$$
$$\vdots$$
$$u_k = u_{k-1} p_1 + v_{k-1} q_2,$$
$$\vdots$$
$$u_n = u_{n-1} p_1 + v_{n-1} q_2.$$

A system of "dual equations" also holds when the u's are interchanged with the corresponding v's and the subscripts 1 and 2 of p and q are interchanged.

PROOF

1. The intersection of E_k and "H appears on the kth toss" is equivalent to the intersection of E_{k-1} and "H appears on the kth toss."

Reason. By definition.

2. The intersection of E_k and "T appears on the kth toss" is equivalent to the intersection of F_{k-1} and "T appears on the kth toss."

Reason. By definition.

3. The probability of the intersection in Statement 1 is equal to $u_{k-1}p_1$.

Reason. Recalling that the Consistency Theorem holds for this game, we note that the probability of the above intersection, which is a union of events determined by the first k tosses, has only to be computed for a game of just k tosses. The probability of the intersection of E_{k-1} and "H appears on the kth toss" is a sum of products of k factors p_1, p_2, q_1, and q_2. Each product contains a (common) factor p_1 corresponding to "H on the kth toss"; and the sum of the remaining factors of the products is, by definition, the probability of E_{k-1}, namely, u_{k-1}.

4. The probability of the intersection in Statement 2 is equal to $v_{k-1}q_2$.

Reason. The reasoning is analogous to that of the previous statement: interchange p_1 and u_{k-1} with q_2 and v_{k-1}, respectively.

5. The asserted equations are valid.

Reason. The equation $u_1 = p_1$ is a simple consequence of the Consistency Theorem. The remaining equations are supported by Statements 1 through 4: E_k is decomposed into two disjoint events whose respective probabilities are given in Statements 3 and 4 (Proposition 5.5). The validity of the dual equations is proved by analogous reasoning.   ◀

As the tosses are successively done, the player shifts from coin I to coin II and back again, and so on, in accordance with the appearance of T. As defined above, u_n is the probability that, starting with coin I, the player is back to coin I after n tosses; v_n is the probability of his being with coin II at the conclusion of n tosses. If the indices 1 and 2 on p and q are interchanged, then v_n is equivalent to the probability that, starting with coin II, the player is with coin I after n tosses. One of the fundamental results of the theory of Markov chains is the "erosion of the effect" of the initial coin over a long sequence of tosses: the limit of the probability of being with coin I for a given initial coin is *independent of the latter*; in other words, the probability is, for a large number of tosses, practically the same for either initial coin. This is the simplest of a class of "long-run" characteristics of Markov chains, known as "ergodic" properties.

The proof of this result is contained in the following two propositions: the first is a statement of the solution of a certain system of linear difference equations; and the second is the identification of the system in Proposition 12.1 as a particular case of the former.

Proposition 12.2 *Let b and c be two real numbers, $b \neq 1$; and let $x_1, \ldots,$ x_n satisfy the system of equations*

$$x_1 = b + c, \qquad x_2 = bx_1 + c,$$
$$x_3 = bx_2 + c, \qquad \ldots, \qquad x_n = bx_{n-1} + c; \tag{12.3}$$

then

$$x_k = \frac{c}{1-b} + b^{k-1}\left[b + c - \frac{c}{1-b}\right], \qquad k = 2, \ldots, n. \tag{12.4}$$

PROOF. This can be verified by direct substitution for the x's in the system of equations. The expression for x_2 given by the simultaneous solution of the first two equations in (12.3) coincides with that given by Eq. (12.4) for $k = 2$: $x_2 = b(b + c) + c$. Suppose that Eq. (12.4) furnishes the correct solution for $x_2, \ldots, x_m$, where $m < n$; then it also furnishes the correct solution for x_{m+1}:

$$x_{m+1} = bx_m + c = b\left[\frac{c}{1-b} + b^{m-1}\left(b + c - \frac{c}{1-b}\right)\right] + c$$
$$= \frac{c}{1-b} + b^m\left[b + c - \frac{c}{1-b}\right].$$

(The reader is asked to check the last algebraic identity in Exercise 7 below.) This shows that Eq. (12.4) is the solution of the system (12.3). ◀

Proposition 12.3 *The probability u_n is given by*

$$u_n = \frac{q_2}{q_1 + q_2} + (p_1 - q_2)^{n-1}\left(p_1 - \frac{q_2}{q_1 + q_2}\right);$$

thus, if n is large, u_n is very close to $q_2/(q_1 + q_2)$: the sequence $u_1, u_2,$ $\ldots, u_n, \ldots$ has the limit $q_2/(q_1 + q_2)$. The sequence $v_1, v_2, \ldots, v_n, \ldots$ has the limit $q_1/(q_1 + q_2)$ because $u_n + v_n = 1$.

PROOF. Put $b = p_1 - q_2$ and $c = q_2$; then, by Proposition 12.1, $u_1, \ldots,$ u_n satisfies the system of equations (12.3), so that u_n is given by Eq. (12.4) with $k = n$. ◀

The proposition implies that the probability that, starting with coin II, the player is with coin I after the nth toss has the same limit $q_2/(q_1 + q_2)$. This probability is obtained from the above expression for v_n by interchanging the subscripts 1 and 2 of the p's and q's. When the indices are so changed, the limit of the v-sequence given by the proposition becomes $q_2/(q_1 + q_2)$.

12.3 EXERCISES

1. Construct the system of probabilities for the two-coin game with four tosses, supposing that coin I is tossed first. Enumerate the outcomes in the events E_1, E_2, E_3, and E_4.

2. Show by computation that the probabilities of events in Exercise 1 satisfy the system of equations in Proposition 12.1.

3. Using the result of Proposition 12.3, find the values of u_5 and u_6 when $p_1 = .8$ and $p_2 = .3$.

4. Show that in the two-coin game the Outcomes of the successive tosses are not, in general, independent: events determined by different tosses are not independent (cf. Chapter 5). (It is sufficient to consider the case $n = 2$.) In what *particular* case are the Outcomes independent?

5. Furnish the details of the proof of the Consistency Theorem for the probability assignments of the two-coin game. The proof in Section 5.2 must be modified as follows: the outcomes are divided into four (instead of two) disjoint classes: those for which the first (the second) coin is used for the nth toss and H (T) appears on it.

6. Write the dual equations in Proposition 12.1.

7. Verify the algebraic identity used in the proof of Proposition 12.2.

chapter 13

APPLICATION OF THE RANDOM WALK
TO SEQUENTIAL TESTING

In the applications of the binomial distribution in Section 2.2 it was shown how this distribution is used in *testing*—for example, the test of the effectiveness of a new medical treatment, or of the percentage of defectives in a batch of manufactured items. In each case the number of "observations" (tosses)—patients, items, etc.—is prescribed before the testing begins; the probability statements about the resulting inferences depend on the adherence to the prescribed number of observations. In many situations—in particular, in medical and industrial testing—each observation is costly; hence, we seek to minimize the number of necessary observations. In this concluding chapter, the theory of "sequential" tests is developed; such tests permit significant reductions in the number of necessary observations. The theoretical basis of such tests represents a useful application of the random walk in Chapter 6. The concept of "stopping time" is used in the comparison of conventional and sequential tests.

13.1 FIXED-SAMPLE-SIZE AND SEQUENTIAL TESTS

Suppose that the probability p of a coin is assumed to be one of exactly two specified values p_0 and p_1, and that we wish to determine which of the two is true for the coin. The fundamental procedure for testing p_0 against p_1 is to toss the coin several times, observe the successive outcomes, and then, on the basis of such observed outcomes, decide which of the two is true. In order to present the theory of testing in its most elementary form, we shall assume throughout this chapter that p_0 and p_1 are related through the equation $p_0 + p_1 = 1$; also, that p_0 and p_1 are not equal (otherwise, the entire problem is meaningless). If $p = p_0$, we say that "the hypothesis p_0 is true"; if $p = p_1$, "the hypothesis p_1 is true." The results that will be derived here are actually valid in much more general situations.

In every testing procedure, there is a possibility of making an incorrect decision: it is possible to decide, on the basis of the observed tosses, that p_1 is true when it is really false, and conversely. With each test we associate two error probabilities: the probability of choosing p_1 when p_0 is true, and the probability of choosing p_0 when p_1 is true. These probabilities may determine which of several tests is used; a test with smaller error probabilities is preferable to one with larger error probabilities.

Here is the first of three different tests of the hypothesis p_0 against the hypothesis p_1. Let p_0 be smaller than p_1. Toss the coin a certain number of times (n): if the Number of H's turns out to exceed the Number of T's, then we accept p_1; and if the Number of H's is less than the Number of T's, then we accept p_0. The case of an equal number of H's and T's can be eliminated from consideration by choosing the number of tosses n to be odd. In this test, the number of tosses is prescribed in advance; in standard statistical terminology, this is called a test with a "fixed sample size."

Example 13.1 Suppose $p_0 = \frac{1}{4}$, $p_1 = \frac{3}{4}$, and $n = 5$. Let us calculate the error probabilities for the above test of p_0 versus p_1. If p_0 is true, the test leads to an incorrect decision if the Number of H's exceeds the Number of T's, or, equivalently, the Number of H's is at least 3. By Definition 2.4, the probability of this event when $p = \frac{1}{4}$ is

$$\binom{5}{3}\left(\frac{1}{4}\right)^3\left(\frac{3}{4}\right)^2 + \binom{5}{4}\left(\frac{1}{4}\right)^4\left(\frac{3}{4}\right) + \binom{5}{5}\left(\frac{1}{4}\right)^5\left(\frac{3}{4}\right)^0 = (90 + 15 + 1)/4^5$$

$$= 99/1024,$$

or approximately .1. If p_1 is true, there is an incorrect decision if the Number of H's is less than 3; by the symmetry of the binomial distribution, the probability of such an error is also .1.   ◁

In the test of fixed sample size we accept p_0 or p_1 accordingly as the fraction of T's or H's is greater than $\frac{1}{2}$. If, at some point before n tosses have actually been completed, the Number of T's or the Number of H's exceeds $n/2$, then it is not necessary to execute the remaining tosses; we can make our decision with fewer than n observations. Such a test is called a "truncated sample" test.

Example 13.2 Assume the same conditions as in Example 13.1: $p_0 = \frac{1}{4}$ and $p_1 = \frac{3}{4}$. Toss the coin until either 3 H's or 3 T's appear: in the former case accept p_1, and in the latter case p_0. At most, five tosses are necessary; however, as few as three may be sufficient. The error probabilities for this test are exactly the same as for the "fixed sample size" test above: every outcome implying acceptance of p_0 or p_1 in the first test also implies the same in the second test.   ◁

This idea can be carried even further: if the first few tosses yield a preponderance of either H's or T's, then we can make a decision without tossing any more. This is the idea behind the notion of the *sequential* test, which is now introduced. The sequential test employs the game of Cain vs. Abel. Let the two players start with equal initial capital $c/2$, where c is even, and let the coin be tossed until either one of the players is ruined or n tosses have been completed. If Abel wins, then we accept p_1; and if Cain wins, then we accept p_0. The probability that neither player is ruined by the nth toss is, in accordance with the Ruin Principle (Section 6.2), very small if n is large; hence, we ignore the possibility that neither player is ruined.

Example 13.3 (Sequential test) Assume again that $p_0 = \frac{1}{4}$ and $p_1 = \frac{3}{4}$; and let us grant Cain and Abel equal initial capital $c/2 = 2$. The hypothesis p_0 or p_1 is accepted accordingly as Cain or Abel wins. Let us calculate the error probabilities for this test: first, the probability that Abel wins (Cain is ruined) when $p = p_0 = \frac{1}{4}$. We apply Proposition 6.6 and compute the probability w_2 from formula (6.8):

$$w_2 = \frac{(\frac{1}{3})^4 - (\frac{1}{3})^2}{(\frac{1}{3})^4 - 1} = .1;$$

hence, the probability of error is approximately equal to that in the test in Example 13.1, where five tosses were prescribed. By symmetry, the probability that Cain wins when $p = p_1 = \frac{3}{4}$, which is the probability that p_0 is accepted when p_1 is true, is also equal to .1.   ◁

The tests described in Examples 13.1, 13.2, and 13.3, respectively, have the same probability of error; which one is best? In many applications not only is the error probability an important criterion in selecting a test, but also the *number of tosses necessary to make the decision*. Let us cite one of the applications in Section 2.2, the prediction of the outcome of an election between two parties, the Blue and Green parties. Consider the simplified case in which it is known that one of the two parties forms three-fourths of the electorate, and that the poll taker wants to determine which of the parties it is. The number of voters sampled corresponds to the number of tosses of the coin. If two different tests have the same error probabilities but require different numbers of voters to be interviewed, then it is reasonable to use the test requiring fewer interviews (or "tosses"), since it is less expensive. The factor of minimizing the number of tosses has clear importance in some of the other applications mentioned: medical testing and industrial acceptance sampling.

Having justified the criterion of number of tosses necessary for a test, we find immediate difficulty in applying this criterion to the comparison of the fixed-sample-size test (Example 13.1), the truncated-sample test (Example 13.2), and the sequential test (Example 13.3): the number of tosses necessary for the last two are random variables; therefore, they cannot be determined before the test is actually made. The number of tosses necessary for the sequential test may, in some cases, turn out to be less than the numbers in the fixed- or truncated-sample tests; on the other hand, in other cases, it may be more. To avoid this ambiguity we shall compare the number of tosses in the fixed-sample-size test to the *expected* number of tosses necessary in the sequential and truncated tests. This comparison is justified by the General Law of Large Numbers (Section 12.1): in repeated use of a sequential test under "identical conditions," the average number of tosses necessary for the tests tends, in the sense of the Law, to be close to the expected number of tosses in a single test.

We point out that, even though a sequential test, as will be shown, usually requires a fewer expected number of tosses than the other two tests, it is not always practical; for example, in the case of a disease with a long period of recovery, it is inadvisable to wait for the outcome in one patient before treating the next one.

13.1 EXERCISES

1. Calculate the probability of error for the test in Example 13.1 for the hypothetical values $p_0 = 2$ and $p_1 = .8$.

2. Repeat Exercise 1 for the test in Example 13.2.

3. Test the hypothesis $p_0 = .3$ against $p_1 = .7$ with $n = 7$ tosses; find the probability of error.

4. Find the probability of error for the sequential test of $p_0 = .3$ against $p_1 = .7$ when the two players are assigned equal initial capital 3.

5. Repeat Exercise 4 for $p_0 = .2$, $p_1 = .8$, and equal initial capital 4.

6. Repeat Exercise 4 for $p_0 = .4$, $p_1 = .6$, and equal initial capital 5.

13.2 STOPPING TIMES

In the game of n tosses played by Cain and Abel, let N be the random variable representing the number of tosses actually made: it is the smaller of n and the number of tosses before one of the players is ruined. This random variable is called the "stopping time" of the game; this terminology is suggested by thinking of the tosses as being performed at regular epochs of time. The distribution of N is easy to calculate for small numbers n.

Example 13.4 Consider a game of four tosses in which Cain and Abel each start with an initial capital of two dollars. Here are the steps in the calculation of the probability distribution of the stopping time N: we first find the probability of each outcome, and then the probabilities of the events $\{N = 2\}$ and $\{N = 4\}$:

outcome	probability	numerical representation
[HHHH]	p^4	2
[HHHT]	p^3q	2
[HHTH]	p^3q	2
[HTHH]	p^3q	4
[THHH]	p^3q	4
[HHTT]	p^2q^2	2
[HTHT]	p^2q^2	4
[HTTH]	p^2q^2	4
[TTHH]	p^2q^2	2
[THTH]	p^2q^2	4
[THHT]	p^2q^2	4
[TTTH]	pq^3	2
[TTHT]	pq^3	2
[THTT]	pq^3	4
[HTTT]	pq^3	4
[TTTT]	q^4	2

It follows that

$$\Pr(N = 2) = p^4 + 2p^3q + 2p^2q^2 + 2pq^3 + q^4,$$
$$\Pr(N = 4) = 2p^3q + 4p^2q^2 + 2pq^3.$$

◁

This kind of computation is very long when n is large, which is just the case in sequential testing. The general theory of stopping times furnishes a simple formula for the expected value of N. This section is devoted to the elements of that theory.

Definition 13.1 The Terminal Fortune is Abel's net gain at the stopping time: it is equal to

a) all of Cain's initial capital if Cain is ruined;
b) the negative of Abel's initial capital if Abel is ruined;
c) Abel's Fortune after the nth toss if neither player is ruined before the nth toss.

Example 13.5 We calculate the probability distribution of the Terminal Fortune in the particular case of $n = 4$ tosses in Example 13.4. The event "Cain is ruined" consists of the outcomes

[HHHH],   [HHHT],   [HHTH],   [HTHH],   [THHH],   [HHTT];

hence, this event has probability $p^4 + 4p^3q + p^2q^2$. The event "Abel is ruined" consists of the outcomes

[TTHH],   [TTTH],   [TTHT],   [THTT],   [HTTT],   [TTTT];

hence, it has probability $q^4 + 4q^3p + q^2p^2$. The Terminal Fortune is 0 if neither player is ruined; this event consists of the outcomes

[HTHT],   [HTTH],   [THTH],   [THHT];

hence, its probability is $4p^2q^2$. In summary, the Terminal Fortune has the following probability distribution:

$$\Pr(\text{Terminal Fortune} = 2) = p^4 + 4p^3q + p^2q^2,$$
$$\Pr(\text{Terminal Fortune} = -2) = q^4 + 4q^3p + p^2q^2,$$
$$\Pr(\text{Terminal Fortune} = 0) = 4p^2q^2. \qquad \lhd$$

The Terminal Fortune will be denoted by S. We now derive the important relation between the probabilities of ruin and the expected value of S.

Proposition 13.1 *Cain and Abel each start with half of the total capital c. Let w be the probability that Cain is ruined. If the scheduled number of tosses n is very large, then the expected value of S is approximately equal to*

$$(c/2)(2w - 1). \qquad (13.1)$$

PROOF

1. The event $\{S = c/2\}$ is identical with the event "Cain is ruined", and $\{S = -c/2\}$ with the event "Abel is ruined"; S assigns values strictly between $-c/2$ and $c/2$ to outcomes in the event "neither is ruined."

Reason. Definition 13.1.

2. If n is large, the probability that Abel is ruined is approximately $1 - w$.

Reason. By the Ruin Principle, the probability that *neither* player is ruined is negligibly small.

3. $E(S)$ is equal to the sum

$$(c/2) \cdot \text{probability that Cain is ruined}$$
$$+ (-c/2) \cdot \text{probability that Abel is ruined}$$
$$+ ((c/2) - 1) \cdot \text{probability of fortune } ((c/2) - 1) \text{ after } n \text{ tosses}$$
$$+ ((c/2) - 2) \cdot \text{probability of fortune } ((c/2) - 2) \text{ after } n \text{ tosses}$$
$$+ \cdots$$
$$+ ((-c/2) + 1) \cdot \text{probability of fortune } ((-c/2) + 1) \text{ after } n \text{ tosses.}$$

Reason. Definitions of expected value (Definition 7.4) and of S (Definition 13.1).

4. The terms after the first two listed in Statement 3 have a sum not exceeding $c/2$ times the probability that neither player is ruined.

Reason. The absolute values of the numerical coefficients $(c/2) - 1$, $(c/2) - 2, \ldots, (-c/2) + 1$ are all smaller than $c/2$; and the sum of the corresponding probabilities is the probability that neither player is ruined.

5. If n is large, $E(S)$ is approximately equal to $(c/2)w + (-c/2)(1 - w)$, which is equal to expression (13.1).

Reason. The first term in the sum in Statement 3 is equal to $(c/2)w$, by definition; the second term is approximately equal to $(-c/2)(1 - w)$, by Statement 2; and, by Statements 1 and 4 and the Ruin Principle, the sum of the remaining terms is negligibly small. ◀

A characteristic property of the stopping time is contained in the following statement:

Proposition 13.2 *For every integer* k, $1 \le k \le n$, *the event* $\{N = k\}$ *is a union of disjoint events determined by the first* k *tosses (see Definition 5.6).*

PROOF. The ruin of one of the players at the kth toss clearly depends *only* on the outcomes of the first k tosses; hence, whether or not an outcome belongs to $\{N = k\}$ depends only on the first k letters in the multiplet describing the outcome.   ◀

The Terminal Fortune S has a representation as the sum of a random number of random variables, as defined in the hypothesis of Proposition 11.11; however, the latter proposition is not applicable to the Terminal Fortune, for reasons soon to be explained. Let $X_1, \ldots, X_n$ be the random variables representing the gains of Abel on the respective tosses; these are independent random variables, each assuming the values $+1$ and -1 with probabilities p and q, respectively. The partial sums of the X's,

$$X_1, \qquad X_1 + X_2, \qquad \ldots, \qquad X_1 + \cdots + X_n,$$

represent the Fortune on the successive tosses; hence, the Terminal Fortune S is the same as X_1 when the stopping time N is 1, it is the same as $X_1 + X_2$ when N is 2, $\ldots$, and is the same as $X_1 + \cdots + X_n$ when N is n; thus, S has the same formal representation as given in Statement 1 of the proof of Proposition 11.11. It turns out that Eq. (11.6) in that proposition is valid in our case, too; however, it is *not* a consequence of that proposition because the stopping time N is *not* independent of the X's, as required by the hypothesis.

We shall prove that Eq. (11.6) of Proposition 11.11 is also valid in the case of the stopping time N; in this case the equation is known as "Wald's equation." [A. Wald (1902–1950) was the founder of sequential-test theory.] To assist the reader, we repeat the representation of the Terminal Fortune S which has already been given in the proof of Proposition 11.11. Let $I_1, \ldots, I_n$ be the following indicator random variables: I_1 assigns 1 to outcomes in $\{N = 1\}$, and 0 to all others; $\ldots$; I_n assigns 1 to outcomes in $\{N = n\}$ and 0 to all others; then

$$S = X_1(I_1 + \cdots + I_n) + X_2(I_2 + \cdots + I_n) + \cdots + X_n I_n. \quad (13.2)$$

[Referring to the representation of S in the proof of Proposition 11.11, we note that in Eq. (13.2) we have used the second of the two forms of the representation, and have used the letter n in place of k.]

Proposition 13.3 *For every pair of indices j and k, $j < k$, the random variables I_j and X_k are independent.*

PROOF. The random variables I_j and X_k have the sets of values $(0, 1)$ and $(-1, +1)$, respectively. In order to prove their independence, one has to prove the independence of the following four pairs of events (Definition 9.1):

$$\{X_k = +1\} \quad \text{and} \quad \{N = j\},$$
$$\{X_k = -1\} \quad \text{and} \quad \{N = j\},$$
$$\{X_k = +1\} \quad \text{and} \quad \{N \neq j\},$$
$$\{X_k = -1\} \quad \text{and} \quad \{N \neq j\}.$$

1. The first two events are independent.

Reason. The term $\{X_k = +1\}$ represents an event determined by the kth toss; $\{N = j\}$ is a union of disjoint events determined by the first j tosses. The Independence Theorem and Proposition 5.7 now imply the independence of the two events.

2. The other pairs are also independent.

Reason. The other pairs are obtained from the first two by interchanging with the complements; hence, the statement follows from Statement 1 and Proposition 5.8.   ◀

Proposition 13.4 (*Wald's equation*)

$$E(S) = E(N)(p - q) \tag{13.3}$$

PROOF

1. If $j < k$, then $E(X_k I_j) = (p - q)\Pr(N = j)$.

Reason. By definition, $E(X_k) = p - q$; by Proposition 11.5, $E(I_j) = \Pr(N = j)$; hence, the Statement follows from the independence of X_k and I_j (Proposition 13.3) and Proposition 11.8.

2. The sum of the indicator random variables, $I_1 + \cdots + I_n$, is 1; in other words, every outcome is assigned the numerical representation 1 by

$$I_1 + \cdots + I_n.$$

Reason. Every outcome belongs to one and only one of the events $\{N = 1\}$, ..., $\{N = n\}$; hence, it is assigned the value 1 by exactly *one* of the corresponding indicator random variables $I_1, \ldots, I_n$, and 0 by all others.

3. The random variable X_k is the same as the random variable

$$X_k(I_1 + \cdots + I_n), \quad 1 \leq k \leq n.$$

Reason. Statement 2.

4. For each k, $2 \leq k \leq n$,

$$E[X_k(I_1 + \cdots + I_{k-1})] = (p - q)(\Pr(N = 1) + \cdots + \Pr(N = k - 1)).$$

Reason. By Proposition 11.4, the expected value on the left-hand side of the above equation is equal to $E(X_k I_1) + \cdots + E(X_k I_{k-1})$; hence, the equation above follows from Statement 1.

5. For each k, $1 \leq k \leq n$,

$$E[X_k(I_k + \cdots + I_n)] = (p - q)(\Pr(N = k) + \cdots + \Pr(N = n)).$$

Reason. In the case $k = 1$, the equation

$$E[X_1(I_1 + \cdots + I_n)] = (p - q)(\Pr(N = 1) + \cdots + \Pr(N = n))$$

follows from Statement 3 and from the fact that the sum of the terms of the probability distribution of N is 1. When k is greater than 1 (and less than or equal to n), we write $E(X_k)$ as

$$E[X_k(I_1 + \cdots + I_{k-1})] + E[X_k(I_k + \cdots + I_n)]$$

(employing Proposition 11.2). Applying Statement 4, we get

$$\begin{aligned} p - q = (p - q)(\Pr(N = 1) + \cdots + \Pr(N = k - 1)) \\ + E[X_k(I_k + \cdots + I_n)]. \end{aligned}$$

We can now derive the assertion of Statement 5 from this equation by noting that

$$\begin{aligned} \Pr(N = 1) + \cdots + \Pr(N = k - 1) \\ = 1 - \Pr(N = k) - \cdots - \Pr(N = n). \end{aligned}$$

6. $\begin{aligned} E(S) = (p - q)[\Pr(N = 1) + \Pr(N = 2) + \cdots + \Pr(N = n)] \\ + (p - q)[\Pr(N = 2) + \cdots + \Pr(N = n)] \\ + (p - q)\Pr(N = n). \end{aligned}$

Reason. By Proposition 11.4, the expected value of the random variable S displayed in Eq. (13.2) is equal to the sum of the expected values given in Statement 5.

7. Equation (13.2) follows from Statement 6.

Reason. We factor $p - q$ from each term on the right-hand side of the equation in Statement 6, and, as in the proof of Proposition 11.11, write the sum of the remaining factors as a sum of sums:

$$\Pr(N = 1) + \Pr(N = 2) + \cdots + \Pr(N = n)$$
$$+ \Pr(N = 2) + \cdots + \Pr(N = n)$$
$$+ \cdots +$$
$$+ \Pr(N = n).$$

Summation by columns yields

$$\Pr(N = 1) + 2 \cdot \Pr(N = 2) + \cdots + n \cdot \Pr(N = n),$$

which, by definition, is equal to $E(N)$.   ◄

As a direct consequence of Propositions 13.1 and 13.4 we have the following equation:

$$E(N) = (c/2)(2w - 1)/(p - q); \tag{13.4}$$

indeed, we equate the values of $E(S)$ in formulas (13.1) and (13.3) and solve for $E(N)$.

13.2 EXERCISES

1. Consider a game of five tosses in which Cain and Abel start each with three dollars.

a) Calculate the distribution of the stopping time.

b) Find the outcomes in the events $\{N = 3\}$ and $\{N = 5\}$. Show that $\{N = 3\}$ is union of disjoint events determined by the first three tosses.

c) Find the distribution of the Terminal Fortune.

2. Repeat parts (a) and (c) of Exercise 1 in the case where each player starts with two dollars. Show that $\{N = 2\}$ and $\{N = 4\}$ are unions of disjoint events determined by the first two and first four tosses, respectively.

3. Verify Wald's equation in the game in Exercise 1 in the particular case $p = .7$.

4. Verify Wald's equation in the game in Exercise 2 in the case $p = .2$.

5. Estimate the error of approximation in using formula (13.1) for $E(S)$ by noting

a) Statement 4 of the proof of Proposition 13.1; and

b) the result of Exercise 9, Section 6.3.

6. Does the proof of Wald's equation (Eq. 13.3) depend on any property of N besides that stated in Proposition 13.2?

7. Express w in terms of c, p, and q by using Eq. (6.8).

8. What is the logical relation between Wald's equation (Eq. 13.3) and Eq. (11.6)? In what ways are the results overlapping or not?

13.3 CONSTRUCTION OF SEQUENTIAL TESTS

In this section we show how to construct a sequential test with a prescribed probability of error w. The illustration in Example 13.3 does not imply such a general procedure: starting with p_0 and p_1, we picked an arbitrary value of c, and then found the error probability from the solution of the ruin equations; thus, w was determined by the chosen value of c. In planning a sequential test, we are *given* the value of w and have to *find* the appropriate value of c; fortunately, there is an explicit formula for c in terms of w:

$$c = 2\,\frac{\log w - \log(1-w)}{\log p - \log q}. \qquad (13.5)$$

The base of the logarithm is unspecified because the fraction in the above ratio is independent of the particular choice of the base. (The proof is assigned as Exercise 1 below.) Now we prove the validity of Eq. (13.5).

> **Proposition 13.5** *If Cain and Abel each begin with half of the total capital c, where c is given by Eq. (13.5), then the probability that Cain is ruined is equal to w.*

PROOF

1. Let Log mean "the logarithm to the base p/q" $\left(y = \text{Log } x \text{ means } x = (p/q)^y \right)$; then, the right-hand side of Eq. (13.5) is equal to

$$2 \text{ Log } \left(w/(1-w) \right).$$

Reason. Recall that $\log A - \log B = \log(A/B)$ and that the logarithm of the base is always 1.

2. For *any* number c the probability that Cain, starting with $c/2$, is ruined is

$$\frac{(p/q)^c - (p/q)^{c/2}}{(p/q)^c - 1}.$$

Reason. Put $r = c/2$ in Eq. (6.8).

3. If $c = 2 \text{ Log } \left(w/(1-w) \right)$, then

$$(p/q)^c = \left(w/(1-w) \right)^2.$$

Reason. We use the identity for logarithms of arbitrary base B: any positive number A may be expressed as the log A power of B.

4. The probability in Statement 2 is equal to w when c is chosen as in Statement 3.

Reason. Substitute $\left(w/(1-w)^{2/c}\right)$ from Statement 3 into the expression in Statement 2, and reduce by algebra (cf. Exercise 2 below). ◀

Example 13.6 Let us illustrate the relation between c and w in Example 13.3; here, $w = .1$, $p = \frac{1}{4}$, and $q = \frac{3}{4}$. For convenience we work with the base 10 logarithm:

$$2\,\frac{\log(.1) - \log(.9)}{\log(\frac{1}{4}) - \log(\frac{3}{4})} = 2\,\frac{\log 9}{\log 3} = 2\,\frac{9542}{4771} = 4.0.$$

(Table IV contains mantissas.) ◁

Example 13.7 Suppose that $p_0 = .2$ and $p_1 = .8$ and that a sequential test with error probability .01 is desired. Equation (13.5) with $w = .01$, $p = .2$, and $q = .8$ provides the appropriate value of c:

$$c = 2\,\frac{\log 99}{\log 4} = 2\,\frac{1.9956}{.6021},$$

or approximately 6.67. Since this is not any even integer, we choose the next such one larger than 6.67, namely, 8. The test is performed by starting each player with four units and stopping as soon as one is ruined. ◁

13.3 EXERCISES

1. Prove that the fraction on the right-hand side of Eq. (13.5) is independent of the choice of the base; use the formula relating the logarithms of a number in two bases. Why is the fraction positive?

2. Perform the algebraic reduction in Statement 4 of the proof.

3. Find the value of c in each of the following cases: $w = .01$, $p = .1$; $w = .05$, $p = .2$; $w = .01$, $p = .4$.

4. Prove: As c gets larger and larger, the probability of error w gets closer and closer to 0.

13.4 EXPECTED NUMBER OF TOSSES
IN TRUNCATED AND SEQUENTIAL TESTS

In this section we compare the expected numbers of tosses in the truncated and sequential tests, respectively. The letter M is used to represent the Number of tosses in the former test, and, as before, N is the stopping time in the latter test.

The formula for $E(M)$ is derived by the method of indicator random variables (Section 11.2). The truncated test is like a "tournament" (Section 7.3): it stops as soon as either the Number of H's or the Number of T's exceeds $n/2$. [Recall that the number n is chosen odd so that $n/2$ is not an integer but $(n + 1)/2$ is.] Another way of stating this is: the test continues beyond the kth toss if and only if the Number of H's and the Number of T's in the first k tosses are both less than $n/2$. Since the Number of H's plus the Number of T's in k tosses is equal to k, the last statement is equivalent to: The test continues beyond the kth toss if and only if the Number of H's in the first k tosses is strictly between

$$k - n/2 \quad \text{and} \quad n/2. \tag{13.6}$$

The truncated test requires at least $(n + 1)/2$ tosses but at most n. Let $I_{(n+1)/2}, I_{(n+1)/2+1}, \ldots, I_{n-1}$ be the following indicator random variables:

$$I_k = 1 \quad \text{for all outcomes in the event that the test}$$
$$\text{continues beyond the } k\text{th toss,}$$

$$= 0 \quad \text{for all outcomes in the complement,}$$

defined for all integers k from $(n + 1)/2$ through $n - 1$, inclusive. The random variable M is expressible in terms of these indicators:

$$M = (n + 1)/2 + I_{(n+1)/2} + \cdots + I_{n-1}; \tag{13.7}$$

indeed, the test terminates at the kth toss if and only if the indicators with subscripts less than k are all equal to 1 and those with subscripts at least k are all equal to 0.

Proposition 13.6 *Let b_k be the probability that the test continues beyond the kth toss, defined for integers k from $(n + 1)/2$ through $n - 1$; then*

$$E(M) = (n + 1)/2 + b_{(n+1)/2} + \cdots + b_{n-1}. \tag{13.8}$$

PROOF

1. $E(I_k) = b_k$.

Reason. The expected value of an indicator is the probability that the indicator is 1 (Proposition 11.5).

2. $E[(n + 1)/2] = (n + 1)/2$.

Reason. The expected value of a constant is the constant (Exercise 5, Section 7.1).

3. The assertion of the proposition follows from Statements 1 and 2.

Reason. The expected value of a sum is the sum of the expected values (Proposition 11.4).   ◀

By virtue of the definition of b_k and the statement leading to formula (13.6), b_k is equal to the sum of the terms

$$\binom{k}{j} p^j q^{k-j}$$

for which the index j is between $k - n/2$ and $n/2$; for example, when $n = 11, k = 7$,

$$b_7 = \binom{7}{2} p^2 q^5 + \binom{7}{3} p^3 q^4 + \binom{7}{4} p^4 q^3 + \binom{7}{5} p^5 q^2.$$

The value of the sum b_k is unchanged when the probabilities p and q are interchanged; this is a consequence of the symmetry of the binomial coefficients (Eq. 1.6).

Example 13.8 Let us calculate $E(M)$ for the truncated test in Example 13.2. The test requires either three, four, or five tosses; here, $(n + 1)/2 = 3$, $n - 1 = 4$, so that the quantities to be computed are

$$b_3 = \binom{3}{1}\left(\frac{3}{4}\right)\left(\frac{1}{4}\right)^2 + \binom{3}{2}\left(\frac{3}{4}\right)^2\left(\frac{1}{4}\right) = \frac{9}{16},$$

$$b_4 = \binom{4}{2}\left(\frac{3}{4}\right)^2\left(\frac{1}{4}\right)^2 = \frac{27}{128}.$$

From Eq. (13.8) we obtain

$$E(M) = 3 + (9/16) + (27/128) = 3.77. \qquad ◁$$

The expression for $E(N)$ is obtained directly from Eq. (13.4). Let w stand for the error probability of the test; it is equal to the probability that Cain is ruined when p (in the random-walk model) is identified with p_0 and q with p_1; thus,

$$E(N) = (c/2)(2w - 1)/(p_0 - p_1). \qquad (13.9)$$

Example 13.9 Calculate $E(N)$ for the sequential test in Example 13.3; here, $c = 4$, $p_0 = \frac{1}{4}$, $p_1 = \frac{3}{4}$, and $w = .1$:

$$E(N) = (4/2)(.2 - 1)/(-.5) = 3.2.$$

The ratio $E(N)/E(M)$ is approximately .84; thus, in this case, the sequential test provides an "expected saving" of approximately 16 percent.   ◁

Now we present an application involving large samples and suitable approximations.

Example 13.10 Two parties form an electorate, the Blue and Green parties. A poll taker tests the hypothesis that the proportion of Blue voters is $p_0 = .45$ against the hypothesis that it is $p_1 = .55$. He wants a test of probability at least .98 of being correct or, equivalently, at most .02 of being incorrect. We shall find the number of voters to be sampled necessary for the fixed-sample-size test and for the sequential test.

For the first test, n voters are sampled at random and p_0 is accepted if the number of Blue voters in the sample is less than $n/2$. The error probability is the probability that more than $n/2$ Blue voters are in the sample when $p = .45$. In accordance with the normal approximation to the binomial (Chapter 4), the probability that the Number of H's in n tosses of a coin with $p = .45$ is at least $(n + 1)/2$ is approximately equal to the area under the standard normal curve from

$$\frac{(n/2) - n(.45)}{(n(.45)(.55))^{1/2}} \quad \text{to} \quad \frac{n - n(.45) + \frac{1}{2}}{(n(.45)(.55))^{1/2}}.$$

If n exceeds 8 (and we may assume this to be so), the second term above exceeds 3. The area to the right of 3 under the standard normal curve is less than .001; hence, we shall ignore it. Therefore, we wish to find a value of n such that the area under the standard normal curve to the right of the first expression above is less than .02. Now from Table II we find that the area to the right of the point 2 is approximately equal to .02; therefore, n is selected so that the first ratio displayed above is at least equal to 2. Simplifying the ratio, we find that

$$\frac{\sqrt{n}\,(.05)}{[(.45)(.55)]^{1/2}} \geq 2;$$

finally, $n \geq 396$. We conclude that the fixed-sample-size test requires a sample of about 396 voters.

Now consider the sequential test. Put $w = .98$; we find c from Eq. (13.5):

$$c = 2\,\frac{\log (.98) - \log (.02)}{\log (.55) - \log (.45)}.$$

For convenience, we use base 10 for the logarithms:

$$c = 2\,\frac{1.9912 - .3010}{1.7404 - 1.6532} = 2(19.4).$$

Convert c to the next even integer, 40. From Eq. (13.4) we get

$$E(N) = (20)(.96/.1) = 192 \text{ (approximately)}.$$

We see that the expected number of voters necessary for this test is only about 48 percent of that for the fixed-sample-size test.

Now we estimate $E(M)$ for the truncated test; unfortunately, the sum in Eq. (13.8) defining $E(M)$ contains many summands, so that an exact evaluation is tedious; thus, we shall be satisfied with determining an approximate lower bound for $E(M)$. As a first step, note that the normal approximation may be used to estimate the quantities b_k defined in Proposition 13.6. By formula (13.6) and the fact that n is odd, b_k is the probability that the Number of H's in k tosses is between $k - (n/2) + \frac{1}{2}$ and $(n/2) - \frac{1}{2}$, inclusive; hence, by the results of Chapter 4, b_k is approximately equal to the area under the standard normal curve from

$$\frac{kq - (n/2)}{(kpq)^{1/2}} \quad \text{to} \quad \frac{(n/2) - kp}{(kpq)^{1/2}}. \tag{13.10}$$

Suppose that the index k satisfies the two inequalities:

$$k \le \frac{n - 3\sqrt{n}}{2q} \quad \text{and} \quad k \le \frac{n - 3\sqrt{n}}{2p}. \tag{13.11}$$

It follows from these and from the inequalities $pq \le \frac{1}{4}$ (Proposition 3.3) and $k < n$ that the first member in formula (13.10) is less than -3 and the second is greater than $+3$ (cf. Exercise 7 below); hence, the corresponding area under the standard normal curve is 1 (to three decimal places); thus, b_k is, with the same accuracy, equal to 1. We conclude from this that *all* the terms b_k of index up to the smaller of

$$\frac{n - 3\sqrt{n}}{2q} \quad \text{and} \quad \frac{n - 3\sqrt{n}}{2p} \tag{13.12}$$

are approximately 1; thus, by Eq. (13.8), $E(M)$ is at least equal to the smaller of the quantities in formula (13.12).

By assumption, p_1 is the larger of p_0 and p_1; thus,

$$(n - 3\sqrt{n})/2p_1$$

is the smaller of the two quantities in formula (13.12); hence, the ratio of $E(M)$ to n is at least

$$(n - 3\sqrt{n})/2np_1. \tag{13.13}$$

In our case ($n = 396$, $p_1 = .55$) the quantity $E(M)$ is *at least* 306; its ratio to 396 is about 77 percent. The ratio of $E(N)$ to $E(M)$ is 192/306, or about .62. ◁

13.4 EXERCISES

1. Consider the test of the hypothesis $p_0 = .4$ against the alternative $p_1 = .6$ with a fixed number $n = 7$ of tosses. Use Table I for (a) and (b).

a) Find the probability of error w.
b) Compute $E(M)$ for the truncated-sample test.
c) Compute c for the sequential test with the same probability of error w.
d) Compute $E(N)$.

2. Repeat Exercise 1 for $p_0 = .3$, $p_1 = .7$, $n = 9$.

3. Repeat Exercise 1 for $p_0 = .2$, $p_1 = .8$, $n = 7$.

4. Suppose we wish to test $p_0 = .4$ against $p_1 = .6$ with prescribed error probability $w = .01$. Find

a) the number of tosses necessary for the fixed-sample-size test;
b) $E(N)$ for the sequential test;
c) a lower bound on $E(M)$ for the truncated test.

5. Repeat Exercise 4 for $p_0 = .4$, $p_1 = .6$, $w = .02$.

6. Repeat Exercise 4 for $p_0 = .48$, $p_1 = .52$, $w = .01$.

7. Prove the statement following formula (13.11): the inequalities

$$k < n, \qquad pq \le \tfrac{1}{4}, \qquad \text{and} \qquad k \le \frac{n - 3\sqrt{n}}{2q}$$

imply

$$\frac{kq - (n/2)}{(kpq)^{1/2}} \le -3;$$

and the inequalities

$$k < n, \qquad pq \leq \tfrac{1}{4}, \qquad \text{and} \qquad k \leq \frac{n - 3\sqrt{n}}{2p}$$

imply

$$\frac{(n/2) - kp}{(kpq)^{1/2}} \geq 3.$$

8. Prove: If the number of tosses n necessary for the fixed-sample-size test is very large, then the lower bound of the ratio of $E(M)$ to n is very nearly equal to $1/2p_1$. [Hint: Show that the latter is the limit of the sequence of numbers in formula (13.13) for $n = 1, 2, \ldots$.]

TABLES

TABLES

Table I   The Binomial Distribution*

n	k	p = .05	p = .1	p = .2	p = .3	p = .4	n	k	p = .05	p = .1	p = .2	p = .3	p = .4
2	0	.9025	.8100	.6400	.4900	.3600	8	0	.6634	.4305	.1678	.0576	.0168
	1	.0950	.1800	.3200	.4200	.4800		1	.2793	.3826	.3355	.1977	.0896
	2	.0025	.0100	.0400	.0900	.1600		2	.0515	.1488	.2936	.2965	.2090
3	0	.8574	.7290	.5120	.3430	.2160		3	.0054	.0331	.1468	.2541	.2787
	1	.1354	.2430	.3840	.4410	.4320		4	.0004	.0046	.0459	.1361	.2322
	2	.0071	.0270	.0960	.1890	.2880		5		.0004	.0092	.0467	.1239
	3	.0001	.0010	.0080	.0270	.0640		6			.0011	.0100	.0413
4	0	.8145	.6561	.4096	.2401	.1296		7			.0001	.0012	.0079
	1	.1715	.2916	.4096	.4116	.3456		8				.0001	.0007
	2	.0135	.0486	.1536	.2646	.3456	9	0	.6302	.3874	.1342	.0404	.0101
	3	.0005	.0036	.0256	.0756	.1536		1	.2985	.3874	.3020	.1556	.0605
	4		.0001	.0016	.0081	.0256		2	.0629	.1722	.3020	.2668	.1612
5	0	.7738	.5905	.3277	.1681	.0778		3	.0077	.0446	.1762	.2668	.2508
	1	.2036	.3280	.4096	.3602	.2592		4	.0006	.0074	.0661	.1715	.2508
	2	.0214	.0729	.2048	.3087	.3456		5		.0008	.0165	.0735	.1672
	3	.0011	.0081	.0512	.1323	.2304		6		.0001	.0028	.0210	.0743
	4		.0005	.0064	.0284	.0768		7			.0003	.0039	.0212
	5			.0003	.0024	.0102		8				.0004	.0035
6	0	.7351	.5314	.2621	.1176	.0467		9					.0003
	1	.2321	.3543	.3932	.3025	.1866	10	0	.5987	.3487	.1074	.0282	.0060
	2	.0305	.0984	.2458	.3241	.3110		1	.3151	.3874	.2684	.1211	.0403
	3	.0021	.0146	.0819	.1852	.2765		2	.0746	.1937	.3020	.2335	.1209
	4	.0001	.0012	.0154	.0595	.1382		3	.0105	.0574	.2013	.2668	.2150
	5		.0001	.0015	.0102	.0369		4	.0010	.0112	.0881	.2001	.2508
	6			.0001	.0007	.0041		5	.0001	.0015	.0264	.1029	.2007
7	0	.6983	.4783	.2097	.0824	.0280		6		.0001	.0055	.0368	.1115
	1	.2573	.3720	.3670	.2471	.1306		7			.0008	.0090	.0425
	2	.0406	.1240	.2753	.3176	.2613		8			.0001	.0014	.0106
	3	.0036	.0230	.1147	.2269	.2903		9				.0001	.0016
	4	.0002	.0026	.0287	.0972	.1935		10					.0001
	5		.0002	.0043	.0250	.0774							
	6			.0004	.0036	.0172							
	7				.0002	.0016							

The probability of k H's in n tosses for a given value of p is equal to the probability of $n - k$ H's in n tosses for the value $1 - p$; thus, the probabilities for $p = .6, .7, \ldots$ are obtained by reversing the probabilities for .4, .3, $\ldots$, respectively; for example, the probability of 2 H's in 6 tosses with $p = .7$ is equal to the probability of 4 H's with $p = .3$, namely, .0595.

* Adapted from J. L. Hodges and E. L. Lehmann, *Elements of Finite Probability*, San Francisco: Holden-Day, Inc., 1965, with permission.

Table I The Binomial Distribution (Continued)

n	k	p = .5	n	k	p = .5	n	k	p = .5	n	k	p = .5	n	k	p = .5
2	0	.2500	13	0	.0001	18	0	.0000	23	2	.0000	27	3	.0000
	1	.5000		1	.0016		1	.0001		3	.0002		4	.0001
3	0	.1250		2	.0095		2	.0006		4	.0011		5	.0006
	1	.3750		3	.0349		3	.0031		5	.0040		6	.0022
4	0	.0625		4	.0873		4	.0117		6	.0120		7	.0066
	1	.2500		5	.1571		5	.0327		7	.0292		8	.0165
	2	.3750		6	.2095		6	.0708		8	.0584		9	.0349
5	0	.0312	14	0	.0001		7	.1214		9	.0974		10	.0629
	1	.1562		1	.0009		8	.1669		10	.1364		11	.0971
	2	.3125		2	.0056		9	.1855		11	.1612		12	.1295
6	0	.0156		3	.0222	19	1	.0000	24	2	.0000		13	.1494
	1	.0938		4	.0611		2	.0003		3	.0001	28	3	.0000
	2	.2344		5	.1222		3	.0018		4	.0006		4	.0001
	3	.3125		6	.1833		4	.0074		5	.0025		5	.0004
7	0	.0078		7	.2095		5	.0222		6	.0080		6	.0014
	1	.0547	15	0	.0000		6	.0518		7	.0206		7	.0044
	2	.1641		1	.0005		7	.0961		8	.0438		8	.0116
	3	.2734		2	.0032		8	.1442		9	.0779		9	.0257
8	0	.0039		3	.0139		9	.1762		10	.1169		10	.0489
	1	.0312		4	.0417	20	1	.0000		11	.1488		11	.0800
	2	.1094		5	.0916		2	.0002		12	.1612		12	.1133
	3	.2188		6	.1527		3	.0011	25	2	.0000		13	.1395
	4	.2734		7	.1964		4	.0046		3	.0001		14	.1494
9	0	.0020	16	0	.0000		5	.0148		4	.0004	29	4	.0000
	1	.0176		1	.0002		6	.0370		5	.0016		5	.0002
	2	.0703		2	.0018		7	.0739		6	.0053		6	.0009
	3	.1641		3	.0085		8	.1201		7	.0143		7	.0029
	4	.2461		4	.0278		9	.1602		8	.0322		8	.0080
10	0	.0010		5	.0667		10	.1762		9	.0609		9	.0187
	1	.0098		6	.1222	21	1	.0000		10	.0974		10	.0373
	2	.0439		7	.1746		2	.0001		11	.1328		11	.0644
	3	.1172		8	.1964		3	.0006		12	.1550		12	.0967
	4	.2051	17	0	.0000		4	.0029	26	3	.0000		13	.1264
	5	.2461		1	.0001		5	.0097		4	.0002		14	.1445
11	0	.0005		2	.0010		6	.0259		5	.0010	30	4	.0000
	1	.0054		3	.0052		7	.0554		6	.0034		5	.0001
	2	.0269		4	.0182		8	.0970		7	.0098		6	.0006
	3	.0806		5	.0472		9	.1402		8	.0233		7	.0019
	4	.1611		6	.0944		10	.1682		9	.0466		8	.0055
	5	.2256		7	.1484	22	1	.0000		10	.0792		9	.0133
12	0	.0002		8	.1855		2	.0001		11	.1151		10	.0280
	1	.0029					3	.0004		12	.1439		11	.0509
	2	.0161					4	.0017		13	.1550		12	.0806
	3	.0537					5	.0063					13	.1115
	4	.1208					6	.0178					14	.1354
	5	.1934					7	.0407					15	.1445
	6	.2256					8	.0762							
							9	.1186						
							10	.1542						
							11	.1682						

By the symmetry of the binomial coefficients, the probability for $k > n/2$ is the same as that for $n - k$; for example, the probability of 4 H's in six tosses is the same as that of 2 H's, namely, .2344.

Table II   Normal Curve Areas*

Area under the standard normal curve
from 0 to z, shown shaded, is $A(z)$.

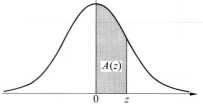

z	.00	.01	.02	.03	.04	.05	.06	.07	.08	.09
0.0	.0000	.0040	.0080	.0120	.0160	.0199	.0239	.0279	.0319	.0359
0.1	.0398	.0438	.0478	.0517	.0557	.0596	.0636	.0675	.0714	.0753
0.2	.0793	.0832	.0871	.0910	.0948	.0987	.1026	.1064	.1103	.1141
0.3	.1179	.1217	.1255	.1293	.1331	.1368	.1406	.1443	.1480	.1517
0.4	.1554	.1591	.1628	.1664	.1700	.1736	.1772	.1808	.1844	.1879
0.5	.1915	.1950	.1985	.2019	.2054	.2088	.2123	.2157	.2190	.2224
0.6	.2257	.2291	.2324	.2357	.2389	.2422	.2454	.2486	.2517	.2549
0.7	.2580	.2611	.2642	.2673	.2704	.2734	.2764	.2794	.2823	.2852
0.8	.2881	.2910	.2939	.2967	.2995	.3023	.3051	.3078	.3106	.3133
0.9	.3159	.3186	.3212	.3238	.3264	.3289	.3315	.3340	.3365	.3389
1.0	.3413	.3438	.3461	.3485	.3508	.3531	.3554	.3577	.3599	.3621
1.1	.3643	.3665	.3686	.3708	.3729	.3749	.3770	.3790	.3810	.3830
1.2	.3849	.3869	.3888	.3907	.3925	.3944	.3962	.3980	.3997	.4015
1.3	.4032	.4049	.4066	.4082	.4099	.4115	.4131	.4147	.4162	.4177
1.4	.4192	.4207	.4222	.4236	.4251	.4265	.4279	.4292	.4306	.4319
1.5	.4332	.4345	.4357	.4370	.4382	.4394	.4406	.4418	.4429	.4441
1.6	.4452	.4463	.4474	.4484	.4495	.4505	.4515	.4525	.4535	.4545
1.7	.4554	.4564	.4573	.4582	.4591	.4599	.4608	.4616	.4625	.4633
1.8	.4641	.4649	.4656	.4664	.4671	.4678	.4686	.4693	.4699	.4706
1.9	.4713	.4719	.4726	.4732	.4738	.4744	.4750	.4756	.4761	.4767
2.0	.4772	.4778	.4783	.4788	.4793	.4798	.4803	.4808	.4812	.4817
2.1	.4821	.4826	.4830	.4834	.4838	.4842	.4846	.4850	.4854	.4857
2.2	.4861	.4864	.4868	.4871	.4875	.4878	.4881	.4884	.4887	.4890
2.3	.4893	.4896	.4898	.4901	.4904	.4906	.4909	.4911	.4913	.4916
2.4	.4918	.4920	.4922	.4925	.4927	.4929	.4931	.4932	.4934	.4936
2.5	.4938	.4940	.4941	.4943	.4945	.4946	.4948	.4949	.4951	.4952
2.6	.4953	.4955	.4956	.4957	.4959	.4960	.4961	.4962	.4963	.4964
2.7	.4965	.4966	.4967	.4968	.4969	.4970	.4971	.4972	.4973	.4974
2.8	.4974	.4975	.4976	.4977	.4977	.4978	.4979	.4979	.4980	.4981
2.9	.4981	.4982	.4982	.4983	.4984	.4984	.4985	.4985	.4986	.4986
3.0	.4987	.4987	.4987	.4988	.4988	.4989	.4989	.4989	.4990	.4990

* From Frederick Mosteller, Robert E. K. Rourke, and George B. Thomas, Jr.,
Probability with Statistical Applications, Reading, Mass.: Addison-Wesley, 1961.

Table III The Poisson Distribution*

k \ λ	1	2	3	4	5	6	7	8	9	10
0	.3679	.1353	.0498	.0183	.0067	.0025	.0009	.0003	.0001	.0000
1	.3679	.2707	.1494	.0733	.0337	.0149	.0064	.0027	.0011	.0005
2	.1839	.2707	.2240	.1465	.0842	.0446	.0223	.0107	.0050	.0023
3	.0613	.1804	.2240	.1954	.1404	.0892	.0521	.0286	.0150	.0076
4	.0153	.0902	.1680	.1954	.1755	.1339	.0912	.0572	.0337	.0189
5	.0031	.0361	.1008	.1563	.1755	.1606	.1277	.0916	.0607	.0378
6	.0005	.0120	.0504	.1042	.1462	.1606	.1490	.1221	.0911	.0631
7	.0001	.0034	.0216	.0595	.1044	.1377	.1490	.1396	.1171	.0901
8		.0009	.0081	.0298	.0653	.1033	.1304	.1396	.1318	.1126
9		.0002	.0027	.0132	.0363	.0688	.1014	.1241	.1318	.1251
10			.0008	.0053	.0181	.0413	.0710	.0993	.1186	.1251
11			.0002	.0019	.0082	.0225	.0452	.0722	.0970	.1137
12			.0001	.0006	.0034	.0113	.0264	.0481	.0728	.0948
13				.0002	.0013	.0052	.0142	.0296	.0504	.0729
14				.0001	.0005	.0022	.0071	.0169	.0324	.0521
15					.0002	.0009	.0033	.0090	.0194	.0347
16						.0003	.0014	.0045	.0109	.0217
17						.0001	.0006	.0021	.0058	.0128
18							.0002	.0009	.0029	.0071
19							.0001	.0004	.0014	.0037
20								.0002	.0006	.0019
21								.0001	.0003	.0009
22									.0001	.0004
23										.0002
24										.0001

* Adapted from J. L. Hodges and E. L. Lehmann, *Elements of Finite Probability*, San Francisco: Holden-Day, Inc., 1965, with permission.

Table IV Common Logarithms of Numbers

N	0	1	2	3	4	5	6	7	8	9
10	0000	0043	0086	0128	0170	0212	0253	0294	0334	0374
11	0414	0453	0492	0531	0569	0607	0645	0682	0719	0755
12	0792	0828	0864	0899	0934	0969	1004	1038	1072	1106
13	1139	1173	1206	1239	1271	1303	1335	1367	1399	1430
14	1461	1492	1523	1553	1584	1614	1644	1673	1703	1732
15	1761	1790	1818	1847	1875	1903	1931	1959	1987	2014
16	2041	2068	2095	2122	2148	2175	2201	2227	2253	2279
17	2304	2330	2355	2380	2405	2430	2455	2480	2504	2529
18	2553	2577	2601	2625	2648	2672	2695	2718	2742	2765
19	2788	2810	2833	2856	2878	2900	2923	2945	2967	2989
20	3010	3032	3054	3075	3096	3118	3139	3160	3181	3201
21	3222	3243	3263	3284	3304	3324	3345	3365	3385	3404
22	3424	3444	3464	3483	3502	3522	3541	3560	3579	3598
23	3617	3636	3655	3674	3692	3711	3729	3747	3766	3784
24	3802	3820	3838	3856	3874	3892	3909	3927	3945	3962
25	3979	3997	4014	4031	4048	4065	4082	4099	4116	4133
26	4150	4166	4183	4200	4216	4232	4249	4265	4281	4298
27	4314	4330	4346	4362	4378	4393	4409	4425	4440	4456
28	4472	4487	4502	4518	4533	4548	4564	4579	4594	4609
29	4624	4639	4654	4669	4683	4698	4713	4728	4742	4757
30	4771	4786	4800	4814	4829	4843	4857	4871	4886	4900
31	4914	4928	4942	4955	4969	4983	4997	5011	5024	5038
32	5051	5065	5079	5092	5105	5119	5132	5145	5159	5172
33	5185	5198	5211	5224	5237	5250	5263	5276	5289	5302
34	5315	5328	5340	5353	5366	5378	5391	5403	5416	5428
35	5441	5453	5465	5478	5490	5502	5514	5527	5539	5551
36	5563	5575	5587	5599	5611	5623	5635	5647	5658	5670
37	5682	5694	5705	5717	5729	5740	5752	5763	5775	5786
38	5798	5809	5821	5832	5843	5855	5866	5877	5888	5899
39	5911	5922	5933	5944	5955	5966	5977	5988	5999	6010
40	6021	6031	6042	6053	6064	6075	6085	6096	6107	6117
41	6128	6138	6149	6160	6170	6180	6191	6201	6212	6222
42	6232	6243	6253	6263	6274	6284	6294	6304	6314	6325
43	6335	6345	6355	6365	6375	6385	6395	6405	6415	6425
44	6435	6444	6454	6464	6474	6484	6493	6503	6513	6522
45	6532	6542	6551	6561	6571	6580	6590	6599	6609	6618
46	6628	6637	6646	6656	6665	6675	6684	6693	6702	6712
47	6721	6730	6739	6749	6758	6767	6776	6785	6794	6803
48	6812	6821	6830	6839	6848	6857	6866	6875	6884	6893
49	6902	6911	6920	6928	6937	6946	6955	6964	6972	6981
50	6990	6998	7007	7016	7024	7033	7042	7050	7059	7067
51	7076	7084	7093	7101	7110	7118	7126	7135	7143	7152
52	7160	7168	7177	7185	7193	7202	7210	7218	7226	7235
53	7243	7251	7259	7267	7275	7284	7292	7300	7308	7316
54	7324	7332	7340	7348	7356	7364	7372	7380	7388	7396
N	0	1	2	3	4	5	6	7	8	9

Table IV   Common Logarithms of Numbers (Continued)

N	0	1	2	3	4	5	6	7	8	9
55	7404	7412	7419	7427	7435	7443	7451	7459	7466	7474
56	7482	7490	7497	7505	7513	7520	7528	7536	7543	7551
57	7559	7566	7574	7582	7589	7597	7604	7612	7619	7627
58	7634	7642	7649	7657	7664	7672	7679	7686	7694	7701
59	7709	7716	7723	7731	7738	7745	7752	7760	7767	7774
60	7782	7789	7796	7803	7810	7818	7825	7832	7839	7846
61	7853	7860	7868	7875	7882	7889	7896	7903	7910	7917
62	7924	7931	7938	7945	7952	7959	7966	7973	7980	7987
63	7993	8000	8007	8014	8021	8028	8035	8041	8048	8055
64	8062	8069	8075	8082	8089	8096	8102	8109	8116	8122
65	8129	8136	8142	8149	8156	8162	8169	8176	8182	8189
66	8195	8202	8209	8215	8222	8228	8235	8241	8248	8254
67	8261	8267	8274	8280	8287	8293	8299	8306	8312	8319
68	8325	8331	8338	8344	8351	8357	8363	8370	8376	8382
69	8388	8395	8401	8407	8414	8420	8426	8432	8439	8445
70	8451	8457	8463	8470	8476	8482	8488	8494	8500	8506
71	8513	8519	8525	8531	8537	8543	8549	8555	8561	8567
72	8573	8579	8585	8591	8597	8603	8609	8615	8621	8627
73	8633	8639	8645	8651	8657	8663	8669	8675	8681	8686
74	8692	8698	8704	8710	8716	8722	8727	8733	8739	8745
75	8751	8756	8762	8768	8774	8779	8785	8791	8797	8802
76	8808	8814	8820	8825	8831	8837	8842	8848	8854	8859
77	8865	8871	8876	8882	8887	8893	8899	8904	8910	8915
78	8921	8927	8932	8938	8943	8949	8954	8960	8965	8971
79	8976	8982	8987	8993	8998	9004	9009	9015	9020	9025
80	9031	9036	9042	9047	9053	9058	9063	9069	9074	9079
81	9085	9090	9096	9101	9106	9112	9117	9122	9128	9133
82	9138	9143	9149	9154	9159	9165	9170	9175	9180	9186
83	9191	9196	9201	9206	9212	9217	9222	9227	9232	9238
84	9243	9248	9253	9258	9263	9269	9274	9279	9284	9289
85	9294	9299	9304	9309	9315	9320	9325	9330	9335	9340
86	9345	9350	9355	9360	9365	9370	9375	9380	9385	9390
87	9395	9400	9405	9410	9415	9420	9425	9430	9435	9440
88	9445	9450	9455	9460	9465	9469	9474	9479	9484	9489
89	9494	9499	9504	9509	9513	9518	9523	9528	9533	9538
90	9542	9547	9552	9557	9562	9566	9571	9576	9581	9586
91	9590	9595	9600	9605	9609	9614	9619	9624	9628	9633
92	9638	9643	9647	9652	9657	9661	9666	9671	9675	9680
93	9685	9689	9694	9699	9703	9708	9713	9717	9722	9727
94	9731	9736	9741	9745	9750	9754	9759	9763	9768	9773
95	9777	9782	9786	9791	9795	9800	9805	9809	9814	9818
96	9823	9827	9832	9836	9841	9845	9850	9854	9859	9863
97	9868	9872	9877	9881	9886	9890	9894	9899	9903	9908
98	9912	9917	9921	9926	9930	9934	9939	9943	9948	9952
99	9956	9961	9965	9969	9974	9978	9983	9987	9991	9996
N	0	1	2	3	4	5	6	7	8	9

ANSWERS TO ODD-NUMBERED EXERCISES

Section 1.1

1. Twenty ordered samples of size two: AB, BA, AC, CA, etc.; two ordered samples obtained from each unordered one

3. Ordered: $100 \cdot 99 \cdot 98 \cdot 97$; unordered: same number divided by 24

Section 1.2

3. 1001; 442; 165

5. $$\binom{N}{s-1} + \binom{N}{s} = \frac{N(N-1)\cdots(N-s+2)}{(s-1)!}$$

$$+ \frac{N(N-1)\cdots(N-s+1)}{s!}$$

$$= \frac{N(N-1)\cdots(N-s+2)}{(s-1)!}\left[1 + \frac{N-s+1}{s}\right]$$

$$= \binom{N+1}{s}$$

7. $\dfrac{\binom{N}{s}}{N^s} = \dfrac{N(N-1)\cdots(N-s+1)}{N^s} = 1\left(1 - \dfrac{1}{N}\right)\cdots\left(1 - \dfrac{s-1}{N}\right);$

in particular case, ratio is .9702

Section 2.1

1.
HHHH	p^4	THHH	p^3q
HHHT	p^3q	HHTT	p^2q^2
HHTH	p^3q	HTHT	p^2q^2
HTHH	p^3q	HTTH	p^2q^2

Eight other outcomes are obtained from these by interchanging H and T, and p and q, respectively.

3. Let A and B correspond to H and T, respectively; then $(A + B)^4$ has summands $AAAA, AAAB, AABA, \ldots$ corresponding to the outcomes HHHH, HHHT, HHTH, $\ldots$, respectively

5. $2^6; 2^8; 2^{10}$

7. $35/128; 231/1024$

9. .4955

11. Use Eq. (1.6)

Section 2.2

1. .8125

3. .3174, .9421

5. a) .1874;   b) .9185

7. p: .1   .2   .3   .4
$L(p)$: .9185 .7373 .5283 .3370

9. .7840

Section 3.1

3. 48; 19.2   5. 7.2; 2.88

7. Interchange p and q, k and $n - k$, and use the symmetry of the binomial coefficients

9. $x = \frac{1}{2}; y = \frac{1}{4}$

Section 3.2

1. a) 3 (Def. 3.5);   b) -3 (Def. 3.5);   c) 4 (Def. 3.5);
   d) 0 (Def. 3.4);   e) 2 (Def. 3.5)

3. If a sequence of x's satisfies the hypothesis of Def. 3.3, it itself serves as the x-sequence in the hypothesis of Def. 3.4; put $L = 0$

7. a) The set has *an* upper bound b; hence, the completeness property implies the existence of a least upper bound
   b) If not, then $L - \epsilon$ would also be an upper bound, smaller than L
   c) Def. 3.5

Section 3.3

1. .9936   3. .9900

5. .9840   7. 9259

9. Follow the proof of Proposition 3.4, using $|k - np|^m$ and $n^m d^m$ in place of $|k - np|^2$ and $n^2 d^2$, respectively

Section 4.1

1. a) .1709;   b) .7887;   c) .0708;
   d) .8023;   e) .1230;   f) .9474
   (Here $n = 200$, $\sqrt{npq} = 6.48$)

3. .0062; .0606; .5867 ($\sqrt{npq} = 15.8$)   5. Each probability is .000

7. 2120.5   9. 2103.7   11. 2627

Section 5.1

1. (See solution to Exercise 1, Section 2.1, for the system of outcomes)

event	outcomes	probability
4 H's	[HHHH]	p^4
3 H's	[HHHT][HHTH]	$4p^3 q$
	[HTHH][THHH]	
2 H's	[HHTT][HTHT][HTTH]	$6p^2 q^2$
	[THTH][TTHH][THHT]	
1 H	[HTTT][THTT]	$4pq^3$
	[TTHT][TTTH]	
0 H's	[TTTT]	q^4

3. (Partial solution)

outcomes	E	F	G	H	(E, H)	(E, F, G)	(F, G)
		sets			unions		intersections
HHHH	x	x			x	x	
HHTH	x	x			x	x	
HTHH	x				x	x	
HTTH	x				x	x	
HHHT		x	x			x	x
HHTT		x	x			x	x
HTHT			x			x	
HTTT			x			x	
TTHH				x	x		
TTHT				x	x		
TTTH				x	x		
TTTT				x	x		

events:	E	F	G	H
probability:	p^2	p^2	pq	q^2

unions:	(E, H)	(E, F, G)	intersection:	(F, G)
probability:	$p^2 + q^2$	p	probability:	$p^2 q$

5. Direct consequence of definitions

7. E_1 = "H, T   on first two tosses, respectively"
   E_2 = "T, H   on first two ..."
   E_3 = "H, H   on first two ..."
   E_4 = "T, T   on first two ..."

   These are disjoint and every outcome belongs to at least one of these.

9.
Intersection	outcomes	probability
E_1, A	[HHH][HHT]	12/27
E_2, A	[THH]	4/27
E_3, A	—	0
E_4, A	—	0
		16/27

Section 5.2

1.
Event	tosses	event	tosses
A	1, 2	E	1, 4
B	3, 4	F	1, 2
C	2, 3	G	1, 4
D	1, 2	H	1, 2

3. If the outcomes specified on the tosses which determine the event are the same, then the events are identical; if not, they are disjoint

Section 5.3

1. a) H on the first and fourth tosses, T on the second;
   b) H on the first toss;   c) sure event;   d) null event

3.
Event	probability	intersection	probability
E_1	pq^2	E_1E_2	p^3q^2
E_2	p^2	E_1E_3	p^2q^3
E_3	pq	E_2E_3	p^3q
		$E_1E_2E_3$	p^4q^3

5. A is a union of disjoint events determined by the first, second and third tosses, respectively; B is a union of disjoint events determined by the fourth, fifth, and sixth tosses; C is a union of disjoint events determined by the seventh, eighth and ninth; the independence of (i) A and B, (ii) A and C, (iii) B and C, (iv) A and the intersection of B and C is a consequence of the Independence Theorem and Proposition 5.7

7. Two (not including the null and sure event); verified by enumerating all events

9. A is a union of the disjoint events: "a tie occurs on the second toss" and "a tie on the fourth but not the second toss"

Section 6.1

1. a) [HHHHH] [HHHTH] [THHHH] [HTHHH]
   [HHHHT] [HHHTT] [THHHT] [TTHHH]

b) [HTTTT]   [TTTTH]   [THTHT]
[THTTT]   [HTHTT]   [THTTH]
[TTHTT]   [HTTHT]   [TTHTH]
[TTTHT]   [HTTTH]   [HTHTH]

c) Complement of the event consisting of
[TTTTT]   [HTTTT]   [TTTTH]

3. a) $p = .8$, $r = 6$, $m = 16$: $1 - (1 - (.8)^6)^{16} = .992$;
   b) $q = .2$, $r = 2$, $m = 50$: $1 - (1 - (.2)^2)^{50} = .870$

5. The proof of Proposition 6.3 depends on Proposition 6.2, which itself depends on the Independence Theorem and Proposition 5.8

Section 6.2

1. i) Consists of [THHH], [THHT] and all outcomes in the event "H on the first toss"
   ii) [TTTT]
   iii) [THTH] [THTT] [TTHH] [TTHT] [TTTH]

3. Cain 1, Abel 3:

event	probability
Cain ruined	$\frac{21}{32}$
Abel ruined	$\frac{5}{32}$
neither ruined	$\frac{6}{32}$

Cain 2, Abel 2:

Cain ruined	$\frac{10}{32}$
Abel ruined	$\frac{10}{32}$
neither ruined	$\frac{12}{32}$

For Cain 3, Abel 1, interchange the names in the first division.

5. By the result of Exercise 4, g_n is less than or equal to $(1 - p^r)^m$; on the other hand, $g_n = 1 - e_n - f_n = (e_0 - e_n) + (f_0 - f_n)$ because $e_0 + f_0 = 1$. Since $(e_0 - e_n)$ and $(f_0 - f_n)$ are both positive and their sum is less than $(1 - p^r)^m$, so is each of them less than $(1 - p^r)^m$.

Section 6.3

1. $w_0 = 1$, $w_1 = qw_2 + p$, $w_2 = qw_3 + pw_1$, $w_3 = qw_4 + pw_2$, $w_4 = pw_3$, $w_5 = 0$
   Solution: $w_1 = .615$, $w_2 = .360$, $w_3 = .189$, $w_4 = .076$

3. Solution: $w_1 = .250$, $w_2 = .062$, $w_3 = .015$, $w_4 = .004$, $w_5 = .001$

5. See outline of proof preceding Example 6.4

7. L is the event that at least one H occurs during the last four tosses, and that it occurs before the appearance of three T's: it consists of these events determined by the last four tosses: H on the second toss and all possible combinations of H and T on the others; T on the second, H on the third, and all other combinations on the fourth and fifth; T on the second and third, H on the fourth, and either H or T on the fifth

9. Put $r = c - 1$, and apply the result of the exercise

Section 7.1

1.

y:	0	1	2	3	4	$E(Y) = \frac{27}{16}$
$\Pr(Y = y)$:	$\frac{1}{16}$	$\frac{7}{16}$	$\frac{5}{16}$	$\frac{2}{16}$	$\frac{1}{16}$	$\text{Var}(Y) = .96$

3.

x:	0	1	2	$E(x) = 2p$
$\Pr(X = x)$:	q^2	$2pq$	p^2	$\text{Var}(X) = 2pq$

5. $\Pr(X = c) = 1$; therefore, $E(X) = c \cdot 1$ and $E(X - c)^2 = (c - c)^2 \cdot 1 = 0$

7. $E(X + c) = $ sum of terms: $(x + c)\Pr(X = x) = $ sum:
$x \cdot \Pr(X = x) + $ sum: $c \cdot \Pr(X = x) = E(X) + c$; $\text{Var}(X + c) = $ sum:
$[(x + c) - E(X + c)]^2\Pr(X = x) = $ sum: $[x - E(X)]^2\Pr(X = x) = \text{Var}(X)$

Section 7.3

1. $\{N = 6\}$ is the union of the disjoint events: "H on the sixth toss, and three H's on the first five," and "T on the sixth toss, and three T's on the first five"; the former has probability $\binom{5}{3}p^4q^2$, the latter $\binom{5}{3}p^2q^4$.

3. $\Pr(N = 5) = p^5 + q^5$; $\Pr(N = 6) = \binom{5}{4}(p^5q + pq^5)$;
   $\Pr(N = 7) = \binom{6}{4}(p^5q^2 + p^2q^5)$; $\Pr(N = 8) = \binom{7}{4}(p^5q^3 + p^3q^5)$;
   $\Pr(N = 9) = \binom{8}{4}p^4q^4$

Section 7.5

1.

$[y, z]$:	[0 1]	[1 1]	[2 1]	[3 1]	[4 −15]
$\Pr(Y = y, Z = z)$:	$\frac{1}{16}$	$\frac{7}{16}$	$\frac{5}{16}$	$\frac{2}{16}$	$\frac{1}{16}$

3.

$[x_1, x_2]$:		[0 0]	[0 1]	[1 0]	[1 1]	[2 0]	[2 1]
$\Pr(X_1 = x_1, X_2 = x_2)$,	$p = .5$:	$\frac{1}{8}$	$\frac{1}{8}$	$\frac{2}{8}$	$\frac{2}{8}$	$\frac{1}{8}$	$\frac{1}{8}$
	$p = .6$:	$8/5^3$	$12/5^3$	$24/5^3$	$36/5^3$	$18/5^3$	$27/5^3$

5. There are 27 values assumed by $[X_1, X_2, X_3]$:

[0, 0, 0], [2, 0, 0], [2, 2, 0], etc.; each has equal probability

7.

$[x, z]$	$\Pr(X = x, Z = z)$	$[x, z]$	$\Pr(X = x, Z = z)$
[0 0]	$\frac{2}{16}$	[3 2]	$\frac{1}{16}$
[0 −2]	$\frac{1}{16}$	[4 4]	$\frac{1}{16}$
[1 0]	$\frac{3}{16}$	[−1 −2]	$\frac{2}{16}$
[1 −2]	$\frac{1}{16}$	[−1 −4]	$\frac{1}{16}$
[2 0]	$\frac{1}{16}$		
[2 2]	$\frac{3}{16}$		

Section 8.1

1. $\frac{4}{35}$; $\frac{18}{35}$; $\frac{12}{35}$; $\frac{1}{35}$ 3. $\frac{3}{15}$; $\frac{9}{15}$; $\frac{3}{15}$

Section 8.2

3. $35/6^5$; $70/6^5$ 5. $55/6^{10}$

Section 8.3

1. M: DE, DF, EF; N: $AD, AE, AF, BD, BE, BF, CD, CE, CF$; union of M, N: all the above outcomes; intersection: null event; $\Pr(M) = \frac{1}{5}$; $\Pr(N) = \frac{9}{15}$;

union has probability $\frac{12}{15}$, intersection 0; Pr(complement M) = $\frac{4}{5}$; Pr(complement N) = $\frac{6}{15}$

3. A: complement of [1 1], [1 2] [2 1]; Pr(A) = $\frac{11}{12}$
   B: complement of [6 5] [5 6] [6 6]; Pr(B) = $\frac{11}{12}$
   C: [2 1] [1 2]; Pr(C) = $\frac{1}{18}$
   Union of A, B is sure event, of probability 1; union of A, C: all outcomes except [1, 1], probability $\frac{35}{36}$; union of B, C: equal to B; intersection of A, B: probability $\frac{5}{6}$; intersection of A, C: probability 0; intersection of A, B, C: probability 0.

Section 8.4

1. Use definition of variance

3. Each of these values has probability $\frac{1}{36}$:

   [1 2] [1 3] [1 4] [1 5] [1 6] [1 7];   [2 3] [2 4] [2 5] [2 6] [2 7] [2 8];   [3 4] ... etc.

5. Pr(X_1 = 1) = .5, Pr(X_1 = 2) = .5; Pr(X_2 = 1) = .4, Pr(X_2 = 2) = .3, Pr(X_2 = 3) = .3; Pr(X_3 = 1) = .2, Pr(X_3 = 2) = .2, Pr(X_3 = 3) = .3, Pr(X_3 = 4) = .3

$[x_1\ x_2]$:	[1 1]	[1 3]	[1 2]	[2 2]	[2 1]	[2 3]
Pr($X_1 = x_1, X_2 = x_2$):	.2	.2	.1	.2	.2	.1

$[x_1\ x_3]$:	[1 1]	[2 2]	[1 3]	[2 4]
Pr($X_1 = x_1, X_3 = x_3$):	.2	.2	.3	.3

$[x_2\ x_3]$:	[1 1]	[2 2]	[3 3]	[1 4]	[2 1]	[3 2]	[1 3]	[2 4]
Pr($X_2 = x_2, X_3 = x_3$):	.1	.1	.2	.2	.1	.1	.1	.1

   $[X_1, X_2, X_3]$ has the probabilities indicated for $t_1, \ldots, t_8$

7. $\frac{7}{8}$

9. Same probability distribution as for sum of Scores except that the sums 2, ..., 12 are replaced by $\frac{2}{2}, \frac{3}{2}, \ldots, \frac{12}{2}$

Section 9.2

1. Not independent

3. Probabilities for:

	$[X_1, X_2]$	$[X_2, X_3]$	$[X_1, X_3]$
[0 0]	$\frac{1}{28}$	0	0
[0 1]	$\frac{4}{28}$	$\frac{3}{28}$	0
[0 2]	$\frac{1}{28}$	$\frac{3}{28}$	$\frac{6}{28}$
[1 0]	$\frac{4}{28}$	$\frac{2}{28}$	0
[1 1]	$\frac{8}{28}$	$\frac{8}{28}$	$\frac{12}{28}$
[1 2]	$\frac{4}{28}$	$\frac{6}{28}$	$\frac{4}{28}$
[2 0]	$\frac{1}{28}$	$\frac{1}{28}$	$\frac{3}{28}$
[2 1]	$\frac{4}{28}$	$\frac{4}{28}$	$\frac{3}{28}$
[2 2]	$\frac{1}{28}$	$\frac{1}{28}$	0

Probabilities for $[X_1, X_2, X_3]$:

[0 1 2]	$\frac{4}{28}$	[1 2 1]	$\frac{4}{28}$
[0 0 2]	$\frac{1}{28}$	[1 0 2]	$\frac{2}{28}$
[1 1 1]	$\frac{6}{28}$	[2 1 1]	$\frac{2}{28}$
[1 0 1]	$\frac{2}{28}$	[2 0 1]	$\frac{1}{28}$
[0 2 2]	$\frac{1}{28}$	[2 2 0]	$\frac{1}{28}$
[2 1 0]	$\frac{2}{28}$	[1 1 2]	$\frac{2}{28}$

None of these are independent.

5. In the given proof of Proposition 9.1 for three random variables, replace the events $\{X = x, Y = y\}$ by $\{X = x, Y = y, W = w\}$

Section 9.5

1. $f_1 = \frac{13}{25}$; $f_2 = \frac{8}{25}$; $f_3 = \frac{4}{25}$

5. Alterations in the statement of the Clustering Principle:

change	to
$X_1, \ldots, X_n$	$[X_1, Y_1], \ldots, [X_n, Y_n]$
$x_1, \ldots, x_k$	$[x_1, y_1], \ldots, [x_k, y_k]$
events $\{X_1 = x\}$	events $\{X_1 = x, Y_1 = y\}$
$\Pr(X_1 = x)$	$\Pr(X_1 = x, Y_1 = y)$

Section 10.1

1. a) $(.6)^3(.4)$;   b) $(.6)^8$;   c) .64

3. The first two statements of the proof are:

$$E(X) = (0)p + (1)pq + (2)pq^2 + (3)pq^3 + (4)pq^4 + 5q^5;$$
$$E(X) = (q - q^2) + 2(q^2 - q^3) + 3(q^3 - q^4) + 4(q^4 - q^5) + 5q^5;$$

apply the distributive law to the last expression

5. The formula for the difference is q^n/p

7. $15   9. .0038; .0680

11.
p:	.02	.04	.06	.1
$L(p)$:	.9810	.8571	.6472	.2650

13. .2236; .4326   15. .3004

Section 11.1

1.
z:	0	1	2	3	4
$\Pr(Z = z)$:	.09	.24	.34	.24	.09

$E(Z) = 2$

3.
x:	0	1	2
$\Pr(X = x)$:	.4	.25	.35

$E(X) = .95$

y:	0	1	2
$\Pr(Y = y)$:	.35	.3	.35

$E(Y) = 1$

z:	0	1	2	3	4
$\Pr(Z = z)$:	.2	.2	.25	.15	.2

$E(Z) = 1.95$

5. $91/3$

7. Put X_i = Number of H's on ith toss; then, Number of H's in n tosses is $X_1 + \cdots + X_n$

Section 11.2

1. $2pq + 6p^2q^2 + 20p^3q^3 + 70p^4q^4$; for $p = .5$: 1.4609

3. $\binom{2}{1}pq + \binom{4}{2}p^2q^2 + \cdots + \binom{2n}{n}p^nq^n$

5. $\{Z = 1\}$ is the union of $\{X_1 = 1\}$ and $\{X_2 = 1\}$, and $\{Z = 0\}$ is the complement of the union

7. a) To each person corresponds a toss of a coin with probability $p = \frac{1}{365}$ of H
   b) $1 - (364/365)^n - (n/365)(364/365)^{n-1}$
   c) As in (b)
   d) Expected value: $365 \times$ quantity (b)

Section 11.3

1.

w:	1	2	3	4	6	9
$\Pr(W = w)$:	.16	.32	.16	.16	.16	.04

$E(W) = 1.8$

3.

w:	0	1	2	4
$\Pr(W = w)$:	.55	.1	.15	.2

5. The event $\{XY = 0\}$ is the union of $\{X = 0\}$ and $\{Y = 0\}$, which is the sure event

7. Put $\{I_j = 1\}$ for outcomes in $\{X = x_j\}$ and 0 for the others

Section 11.4

1. $\{X_1 = 1\}$ is "exactly one H on first two tosses"; $\{I_1 = 0\}$ is the complement of "exactly one H on second two tosses"; $\{X_1 = 1\}$ is the union of the events "H, T on the first and second tosses, respectively" and "T, H on the first and second tosses, respectively"; the complement of $\{I_1 = 0\}$ is the union of two similar events determined by the third and fourth tosses; the independence follows from Independence Theorem and Propositions 5.7 and 5.8

3. Modify Statement 5 of the proof

Section 11.5

1. Statement 1 of the given proof remains unchanged; in the subsequent statements make the changes: $E(Y_1) = \mu_1$, $E(X_n) = \mu_n E(X_{n-1})$

3. i) $\mu < 1$; the numbers μ_n are ultimately less than some number $\mu + \epsilon < 1$, so that the limit of the product is 0
   ii) $\mu > 1$; the numbers μ_n are ultimately greater than some number $\mu - \epsilon > 1$, so that the product increases rapidly with n

Section 12.1

1. Average: $1(\frac{13}{25}) + 2(\frac{8}{25}) + 3(\frac{4}{25})$

3. $E(X) = 7$, $k = 11$, $|x_1| + \cdots + |x_{11}| = 77$; the probability that the average is between $7 - d$ and $7 + d$ is at least $1 - (11)(77)^2/4nd^2$

5. $E(X) = .4$, $k = 3$, $|x_1| + |x_2| + |x_3| = 2$; the probability that the average is between $.4 - d$ and $.4 + d$ is at least $1 - 3/nd^2$

Section 12.2

1. $\frac{6}{7}$; $\frac{4}{58}$

3. The proportions of inoculated persons in the diseased and healthy groups must be compared

Section 12.3

1.
Outcome	probability	outcome	probability
HHHH	p_1^4	HTTH	$p_1^2 q_1 q_2$
HHHT	$p_1^3 q_1$	HTHT	$p_1 q_1 p_2 q_2$
HHTH	$p_1^2 q_1 p_2$	HHTT	$p_1^2 q_1 q_2$
HTHH	$p_1 q_1 p_2^2$	TTTH	$q_1^2 q_2 p_2$
THHH	$q_1 p_2^3$	TTHT	$q_1^2 q_2 p_1$
TTHH	$q_1 q_2 p_1^2$	THTT	$q_1^2 q_2 p_2$
THTH	$q_1 p_2 q_2 p_1$	HTTT	$q_1^2 q_2 p_1$
THHT	$q_1 p_2^2 q_2$	TTTT	$q_1^2 q_2^2$

$E_1 = $ "H on first toss"; $E_2 = $ outcomes starting with HH or TT; $E_3 = $ outcomes with exactly one H or three H's in first three tosses; $E_4 = $ outcomes with exactly zero, two, or four H's

3. $u_5 = (\frac{7}{9}) + (.0001/45)$; $u_6 = (\frac{7}{9}) + (.00001/45)$

5. Statements of modified proof:
1 and 2 are the same;
3. The event in Statement 2 is the union of two disjoint events—E_1: H on the first toss and coin I used for the nth; E_2: H on the first and coin II for the nth;
4. Same as Statement 3 in the original proof;
5. The probabilities of the n-letter outcomes in Statement 4 obtained by adjoining H or T to an $(n - 1)$-letter outcome are equal to the probability of the latter outcome multiplied by p_1, q_1, p_2, or q_2, respectively, depending on the membership of the outcome in E_1 or E_2.

The remaining statements are the same as in the original proof; in the reasons, p and q are appropriately replaced by p_1 and q_1 for outcomes in E_1, and by p_2 and q_2 for outcomes in E_2.

7. By elementary algebra

Section 13.1

1. .0579   3. .1260   5. .0039

Section 13.2

1. a) $\Pr(N = 3) = p^3 + q^3$; $\Pr(N = 5) = 1 - p^3 - q^3$;
b) $\{N = 3\}$ is the union of "H on the first three tosses" and "T on the first three tosses"; $\{N = 5\}$ is the complement of this union;

c) s $\Pr(S = s)$

s	$\Pr(S = s)$
-3	$q^3 + 3q^4 p$
-1	$9q^3 p^2$
$+1$	$9p^3 q^2$
$+3$	$p^3 + 3p^4 q$

3. $E(N) = 4.260$; $E(S) = 1.704$

5. $(c/2)(1 - p^{c-1})^m$, where m is the largest multiple of $c - 1$ not exceeding n

7. Put $r = c/2$ and $w_r = w$

Section 13.3

1. Use the identity: $\log_B x = (\log_A x)(\log_B A)$; fraction positive because $p < q$ equivalent to $w < 1 - w$

3. 4.2, 4.2, 22.7; convert each to next larger even integer

Section 13.4

1. a) .2897;

   b) $E(M) = 4[.1296 + .0256] + 5[(.4)(.1536) + (.6)(.3456)]$
$$+ 6[(.4)(.2304) + (.6)(.3456)] + 7(.2765) = 5.6974;$$

   c) $c = 4.4$; convert to 6;

   d) $E(N) = .6309$ [although $E(N)$ exceeds $E(M)$, the test with $c = 6$ actually has probability of error smaller than w]

3. $w = .0334$, $E(M) = 4.9265$, $c = 4.86$;   convert to 6; $E(N) = 4.667$

5. $n = 102$; $E(N) = 48$; $E(M) = 91$

7. The inequality $kq \le (\frac{1}{2})(n - 3\sqrt{n})$ implies

$$\frac{kq - (n/2)}{(kpq)^{1/2}} \le \frac{-3\sqrt{n}}{2(kpq)^{1/2}} = \frac{-3\sqrt{n/k}}{2\sqrt{pq}};$$

since the last fraction is negative, the inequalities $pq \le \frac{1}{4}$, $k \le n$ imply that the fraction is at most equal to -3; the inequality $kp \le (\frac{1}{2})(n - 3\sqrt{n})$ implies that

$$\frac{(n/2) - kp}{(kpq)^{1/2}} \ge \frac{3\sqrt{n}}{2(kpq)^{1/2}},$$

which is at least equal to $+3$.

INDEX

INDEX

Important Theorems

Definitions